SPACESHIP EARTH IN THE ENVIRONMENTAL AGE, 1960–1990

HISTORY AND PHILOSOPHY OF TECHNOSCIENCE

Series Editor: Alfred Nordmann

TITLES IN THIS SERIES

FORTHCOMING TITLES

SPACESHIP EARTH IN THE ENVIRONMENTAL AGE, 1960–1990

BY

Sabine Höhler

Routledge
Taylor & Francis Group

LONDON AND NEW YORK

First published 2015 by Pickering & Chatto (Publishers) Limited

2 Park Square, Milton Park, Abingdon, Oxon OX14 4RN
711 Third Avenue, New York, NY 10017, USA

Routledge is an imprint of the Taylor & Francis Group, an informa business

First issued in paperback 2016

BRITISH LIBRARY CATALOGUING IN PUBLICATION DATA

Hohler, Sabine author.
Spaceship Earth in the environmental age, 1960–1990. – (History and philosophy of technoscience)
1. Human beings – Forecasting – History – 20th century. 2. Natural resources – Forecasting –History – 20th century. 3. Space flight – Forecasting – History – 20th century.
I. Title II. Series. 301-dc23

ISBN-13: 978-1-8489-3509-9 (hbk)
ISBN-13: 978-1-138-71091-7 (pbk)

Typeset by Pickering & Chatto (Publishers) Limited

CONTENTS

ACKNOWLEDGEMENTS

This book results from a long journey that began when I was a postdoctoral student in Berlin. In 2002 I took the opportunity to wander from the history of science track and take a position in interdisciplinary sustainability research at the Hamburg School of Economics and Politics. I am immensely thankful to my Hamburg colleagues in ecological economics, cultural geography and political science, Fred Luks, Sybille Bauriedl, Matthias Winkler and Delia Schindler, for trusting me to have something meaningful to add to sustainability studies, and for introducing me to a new field. Originally trained as a physicist I thought I could handle interdisciplinarity. But it was with the Hamburg project that I gathered first hands-on experiences and entered into arguments that shook the core of what I believed was good and right. I thank Sabina Gorrissen-Salazar for her confidence and humour in putting up with us.

The Hamburg position asked for a research topic that contributed to the interdisciplinary project and still was able to pass as historical research. The focus on economies of nature was informed by spatial theory, and the questions of how limits are met by and set by material flow analysis and accounting practices in sustainability science instigated me to work on the history of the biosphere in the twentieth century. However, the ecological systems perspective on the biosphere soon felt like a trap. The systems views I dealt with threatened to devour and metabolize any critical position. Moving Spaceship Earth into the centre of the work solved the problem. The figure of the spaceship that denoted earth in so much of the 1960s and 1970s literature allowed for a cultural history perspective. Through naratives of the ship I could connect my earlier work on oceanography and aviation in the nineteenth and early twentieth century to ecology and human ecology in the late twentieth century. Whether airships, ocean-going research vessels or spaceships, whether science fact or science fiction: the ship is at the heart of Western culture, symbolizing containment, refuge, exploration and transition.

This was one of the most challenging and exciting expeditions I ever embarked on. Environmental concerns had just returned to the international agenda under the banner of global climate change, and the environmental humanities were in a formative period. How to qualify with interdisciplinary work in disciplinary structures, however, was still an open question. I very much appreciated the institutional home that TU Darmstadt and the Department of Social and Historical Sciences offered me after Hamburg. Mikael Hård supported and encouraged me most ardently on my way. I received shelter also from other colleagues. Helmuth Trischler hosted me as a scholar in residence at the Deutsches Museum; David Gugerli at ETH Zurich took me into his team at the TG (Technikgeschichte). Patrick Kupper, Lea Haller and Andrea Westermann were fabulous teammates and collaborators. Christoph Mauch offered me an environmental history fellowship at the German Historical Institute in Washington, DC, where I made my first contact with scholars who actually identified as environmental historians. To all these colleagues I would like to extend my appreciation.

Michael Hagner and his history of science group I thank for providing a temporary home at ETH Zurich through most of 2010 and 2011. Ariane Tanner has been an inspiring collaborator ever since. Last but not least I would like to thank my colleagues at KTH Royal Institute of Technology in Stockholm. I have settled in with the Division of History of Science, Technology and Environment, the name of which only partly expresses the programmatic and wholly undisciplined ways of engaging with the techno-geo-enviro-politics of twentieth-century nature. From my present-day perspective as part of a community involved in human-environment relations beyond the traditional claims and stakes of academia, my journey seems like an evident and long overdue undertaking.

Many others were wonderful travel companions. Susanne Bauer, Jens Lachmund, Michelle Murphy, Iris Schröder and Rafael Ziegler were colleagues who became friends, and friends who became colleagues. I cannot mention them all but would like to express my gratitude. My dearest friends and family always were a safe haven. My special thanks go to Anne Teresiak who helped me to navigate the deep waters on the last stretches. On a final note, I would like to thank Alfred Nordmann for suggesting Pickering & Chatto's series History and Philosophy of Technoscience for publication. I am indebted to three encouraging anonymous reviewers and to Philip Good, Sophie Rudland and Stephina Clarke for their tireless engagement with the process to turn the work at long last into a printed book.

LIST OF FIGURES

1 CAPACITY: ENVIRONMENT IN A CENTURY OF SPACE

Et Si Le Vrai Luxe, C'Était L'Espace?

In December 2002 the automobile manufacturer Renault launched a television and print campaign to advertise its new generation of large-capacity limousine, the Espace. The image series *La foule* (the crowd) showed pedestrians moving in metropolitan streets through thick human swarms. All along, however, the pedestrians remain exceptional, surrounded and protected by a comfortable clear space that assumes the form of an Espace. After the camera has followed them on their way, the pedestrians finally climb into their actual Espace and drive out of the frame. The subsequent slogan suggestively asks: *Et si le vrai luxe, c'était l'espace?* The pun translates only loosely into the English phrase 'What if the true luxury were space? The new Renault Espace'.

In advertising automobiles, speed, freedom, independence and overall individual mobility have been replaced more and more with the safety and silence, comfort and cleanliness of their passenger compartments. In this regard the message of the advert is familiar. More intriguing is Renault's storyline that states that in the midst of growing modern urban populations and ensuing spatial constraints personal breathing and moving space have become luxury goods. Renault's commercial promises that the extravagance of perfect privacy and protection can be regained and consummated by means of a modern vehicle, the 'monospace', a technologically controlled and optimized personal environment. Novel in Renault's advertisement is the topic of space itself.

I suggest that the spot promoting the Espace, when released at the outset of the new millennium, underlined a historical process of spatial confinement and control that was driven by the permeation, demarcation and distribution of *geospace* in the twentieth century. At the end of the nineteenth century, a seemingly endless process of global expansion came to a close. In the twentieth century the earth came into view in novel ways, and the notion of living within close natural and political limits took on a new quality and urgency. Concerns about the world were supplemented by a focus on the planet earth as a whole. At the peak of the Cold War, the dynamics of global expansion were halted in a strangely stable state that resulted from the strategic parity between the two superpowers, from their technologies for close mutual observation and from a constant threat of mutual destruction. As the globe no longer seemed to tolerate remoteness, the technological rivalry of expansion into outer space began. In turn, it was extraterrestrial photography that supplied pictures of the 'Blue Marble' in the 1960s and for the first time presented planet earth in its entirety. The photographs from space zoomed in on life's fundamental conditions, which were believed to be unique within the universe. Reaching their climax in the early 1970s, concerns about rising pressures on the environment arose in combination with warnings of the limitations of resource-intensive industrial production and of worldwide population increase. In 1972, the Club of Rome's study *The Limits to Growth* warned against 'the predicament of mankind', which was economical growth and ecological destitution. In the same year, the UN Conference on the Human Environment in Stockholm, the first in a series of so-called earth summits, placed the rising awareness of environmental pollution and resource depletion on the international agenda as a problem of global development. The very word 'environment' became a synonym of crisis and urgency.[1]

'What if the true luxury were space?' Renault's bold suggestion seems to reconfirm an argument that the French historian and philosopher Michel Foucault had brought forth in 1967 when speaking of the twentieth century as the 'epoch of space'.[2] Foucault referred to the ways contemporaries conceptualized limited *geospace* as well as increasingly limited global *biospace* to enclose, display and ultimately manage what was considered ever more unique and fragile on earth: inhabitable environment, life and living conditions. It may be no coincidence that at the same time a figure of speech gained currency that combined the notions of global *geospace* and *biospace*: 'Spaceship Earth'. Spaceship Earth became fundamental in articulating spatial limitations at the beginning of what is sometimes termed the Environmental Revolution. On the one hand, this discursive figure expressed the threat of absolute earthly limits and of a questionable future of planet earth and its inhabitants. On the other hand, Spaceship Earth framed the planet in technoscientific terms and recreated the planet as a new hybrid entity. The figure discursively linked the notion of an ultimately finite living space to the technoscientific solutions and visions of space flight. From the time of its invention and diffusion in the 1960s to its demise in the 1980s, Spaceship Earth combined the fear of limited possibilities of expansion with the anticipated possibilities of modern functional spaces that would be scientifically and technologically maintained.

This book argues that Spaceship Earth not only illustrated but also created a fundamental shift in the conception of life and living space on the earth that brought about new regimes and visions of efficiency. Spaceship Earth signified the threat to earth as a natural human habitat, but it also created expectations for science and technology to provide a 'blueprint for survival',[3] substituting the biosphere of the earth with possible surrogate spaces elsewhere. The book focuses on the singular historical constellation around the year 1970 that may be characterized by the intersection of the aspirations of space flight, an overall obsession with the future, rising environmental concerns, Cold War conflicts, the consciousness of a new global interdependence, and last but not least the hitherto unprecedented potential for intervention – and destruction – by scientific and technological means. The conditions of possibility for this historical situation to emerge were met during the Second World War and in the post-war period; however, in several aspects, this situation was prepared by developments in the late nineteenth and early twentieth centuries. New in the 1960s were not the environmental concerns but the optimistic ideas of being able to turn the dismal fate of the planet into a bright future planned by scientists and engineers. During its lifetime up to the late 1980s when globalization and sustainability replaced planetary maintenance, Spaceship Earth did not simply serve as a metaphor to express the fragility of the planet; rather, Spaceship Earth configured a range of aspects associated with the constraint and crowdedness of earth. Spaceship Earth became the central part of a mythology to present the problems of planetary closure meaningfully and to propose strategies and solutions of escape.

The Age of Capacity

Foucault's insistence to take seriously not only the historical meanings of space, but also spatiality itself as a historical problem was brought forward in a singular historical situation that can neither be captured by terms like the Cold War period, the Space Age or the environmental movement alone, nor by what has come to be called globalization. Rather, the historical situation in question points to a new awareness and significance of the global, of 'globality' around 1970 to which every one of the mentioned singularities contributed, pointing to the new dimensions of humankind's impact on the earth.[4]

These dimensions were literally spatial. In order to assess the meaning of globality adequately we need to explore how the emerging notions of a global community and the projects to inventory and allot the global commons, correlated with the reorganization of global spatial relations.[5] Foucault argued for a historiography considering spaces not as fixed and stable, as preformed containers to be filled with cultural meaning, but as products of historically specific material-semiotic representations. Foucault understood space not as homogeneous, metric and isotropic, but rather as an arrangement of positions and relations; he proposed to comprehend spaces and places as changeable and relational structures, as contingent patterns or textures of dynamically combining and recombining sites and positions. Foucault was one of the early proponents of the 'spatial turn' which in the 1990s opened up a whole range of cultural approaches to explore the spatial characteristics of social and cultural relations.[6]

Foucault propagated a concept of space that dismissed uniform Euclidian space. Moreover, he linked this concept specifically to the major themes of the twentieth century. He pointed to the interaction of places, the shifting neighbourhoods and the configurations of simultaneity following from the redistributions of proximity and distance that modern technologies of acceleration, of mobilization, and of international communication, transport and standardization had brought forward. Furthermore, he was concerned with the specific social and cultural conflicts, which became virulent in the late nineteenth century when the globe was first experienced as noticeably confined. This book develops Foucault's observation 'that the anxiety of our era has to do fundamentally with space'.[7] I examine and elaborate the tentative line of reasoning that the future of humankind would increasingly be linked with knowledges of spatial order and that questions of location, situation and mutual spatial relations would become the quandary of the twentieth century.

'But is there anything about the 1970s – as opposed, say, to the 1920s or 1870s – that should make this the decade in which limits to growth become apparent?'[8] The essentialist answer to this question brought forth by physicist John Holdren and biologist and human ecologist Paul Ehrlich in 1974 would involve the familiar and accepted catastrophes of the so-called environmental decade, population growth, industrialization, dwindling resources and unprecedented pollution. If we try to resist this answer then the question directs us towards a more fundamental conceptual transition: Up to the mid-twentieth century, limits, boundaries and spaces had been *territorially* defined. In the form of territories, spaces and places became the objects of national and international struggles and related to ambitions of world power. Countless studies – especially on, say, the 1920s or 1870s – have explored the territorial endeavours of the Imperial Age and its colonial projects, and abundant work has focused on technologies as part of imperial projects and on the scientific bodies of accumulated data that were transformed into tables, graphs and maps and acquired territorial shape.[9] Historian Charles Maier has convincingly argued for discerning the era of high imperialism between roughly 1860 and 1970 as the 'Age of Territoriality'.[10]

Holdren's and Ehrlich's question points to a curious circumstance: The obsession with territory in history and historiography does not carry all the way through when trying to assess the state of the world around 1970. Undeniably, the national and transnational exploration and occupation of the world by the major imperial powers involved massive territorial boundary work to cause the 'white spots' to gradually disappear from the world map. The African continent, the American West, the earth's poles, the oceans, the atmosphere: having long been *terrae incognitae*, these spaces were mapped, distributed and colonized to a large extent by the First World War. In the same course, the long 'lost horizon'[11] of expansionist projects approached, and the 'World Frontier'[12] came to a close. The very 'cage' of the inhabited earth had been gauged, as the French geographer Jean Brunhes memorably noted on the occasion of his inaugural lecture in 1911.[13] It seems, however, that the limitedness of the earth came into clear focus only after the Second World War when the division of the world into two politically opposed hemispheres intersected with the long-standing separation of the globe into two hemispheres of unequal development and prosperity. I argue that the 1970s are known as the decade in which limits to growth became apparent not so much because of the struggle over political and territorial boundaries as because of a discourse of global limits. I suggest discussing the transition to global awareness around 1970 as a problem of *capacity*.

By the term 'Age of Capacity' I indicate the shift of perspective towards limited global space and global environment around 1970 and, consequently, the focal shift towards the world as a community sharing a common destiny. This move from national territory to global capacity went along with new concepts of 'living space' positioned at the intersection of biopolitical and scientific-technological spatial regimes. The discourse on living space needs to be carefully distinguished from earlier political philosophies and economies of living space related to territory. The debate on living space around 1970 sprang from the rising environmental discourse; living space referred to the rediscovery of the biosphere as a limited spherical container sustaining all life on earth. Moreover, the ecological sciences reframed the biosphere in terms of a closed and complex ecosystem. It is this new and specific meaning of a global living space that I aim to capture when referring to the ecological term of a human habitat.[14]

This work sets out to explore the regimes of efficiency that developed within the broader discourse of Spaceship Earth. I project my topic primarily from a historical science studies perspective, using material from three fields of twentieth-century science: first, systems ecology and its reinvention of nature and environment as a 'life support system'; second, human ecology and its allocation of human beings to available earthly space that became manifest in the concept of an ecological limit of the earth system, the earth's 'carrying capacity'; and third, the intersections of ecology and the science and technology of space flight in the visions to overcome earthly limits by eventually constructing biosphere surrogates and recovering living space elsewhere.

Notwithstanding my approach from the perspective of science and technology studies, the figure of Spaceship Earth emerged and travelled in Western culture in a broad sense, of which the environmental sciences formed only a part, albeit an important one. The discourse of Spaceship Earth profited from several characteristics of the time in question: the booming optimism of the period after the Second World War, with its hope for scientific and technological progress; the perception of growing global interdependencies during the Cold War; and the debate about environmental pollution, resource depletion and population growth that formed part of the rising environmental movement.

A Shrinking Globe

Political interdependence and technoscientific intervention set within an enclosed and finite space are the key features of Spaceship Earth. I would like to discuss these key features as they were expressed in a striking emblem of the 1960s: Unisphere, the largest model of the globe ever built, was the centrepiece and official symbol of the World's Fair that took place in New York in 1964 and 1965 (see Figure 1.1). Unisphere was located on the US Federal and State Area at the Fountain of the Continents and was presented to the world as the symbol of 'Man's Achievements on a Shrinking Globe in an Expanding Universe'.[15]

Unisphere echoed some of the most discussed and often controversial topics of its time. For one, the gigantic steel construction radiated the optimism of the Western world in the prevalent scientific and technological achievements during the late 1950s and early 1960s, in the same way as the futuristic design and architecture of the entire exhibition did. Modernist visions, fashions and architectures as well as the celebration of conveniences and the comfortable lifestyle they promised featured strongly at the New York World's Fair. By exhibiting scientific competence and technological skill, the fair reiterated the high hopes and expectations of the post-war period that scientific and technological advancement would arrive at the answers to some of the most pressing issues in the world. Accordingly, one of the major themes of the New York Fair was 'A Millennium of Progress', a motto that had similarly defined the World's Fair in Chicago in 1933 and was revived after the Second World War.[16]

Second, Unisphere represented the Space Age, the dynamic and expectant decade of space exploration in the 1960s. Centred in a large, circular reflecting pool and surrounded by a series of water-jet fountains designed to obscure its tripod pedestal, Unisphere appeared to the visitors like the earth floating in deep space. The model was constructed to the dimensions at which an astronaut would view the earth from a distance of 6,000 miles in space. The three rings surrounding the model symbolized the orbits of Yuri Gagarin, the Russian cosmonaut and first human in space in 1961, of John Glenn, the first American astronaut who fully circled the earth in 1962, and of Telstar, the first man-made communications satellite that was launched in 1962 by the American telecommunications company AT&T.

Unisphere presented by (USS) United States Steel
© 1961 New York World's Fair 1964-1965 Corporation

Figure 1.1: Unisphere. *New York World's Fair 1964/1965. Official Guide*, **by the editors of** *Time-Life Books* **(New York: Time Inc., 1964), p. 178.**

The orbital rings paid tribute to man's quest to enter space and open up new and unforeseen worlds. They represented the ability to assess the earth as a whole by means of space exploration. Moreover, they symbolized the novelty of global telecommunication through the coupling of radio and television broadcasting and satellite technology.[17]

Thirdly, Unisphere stood for the growing global interdependency that the world experienced in the second half of the twentieth century. The model displayed a novel unity, which despite, and because of, the ongoing international conflicts emphasized a global community slowly gaining contours. Resonating with the vivid memory of the atomic bombs of 1945 and the imminent threat of a nuclear war that was feared to be global in scope and total in its destructive power, the theme of universality again addressed the demonstrated potential of twentieth-century science and technology to change the world's course, for better or for worse. Corresponding to these ideas, the second theme of the New York World's Fair was chosen to be 'Peace through Understanding'. At the peak of the Cold War the motto called for comprehension and appreciation, for a mutual advance and accommodation of the different parts of the world. Unisphere presented the world as a closed sphere, almost a cage in Brunhes's sense, which was to contain everything and everybody, and which did not hold any 'outside' apart from the tentative triumphs of space travel to elude gravity. Unisphere conveyed the message that on the newly discovered planet earth, nothing would be left to put aside, to save or to conquer. Unisphere anticipated the idea of a global commons, of shared common resources and of the need to internationally institutionalize the modes of their management. In relating the delicate wealth of the small earthly world to the vastness of planetary space, Unisphere clearly outlined a shrinking globe.

Unisphere can be seen as a symbol of rising global consciousness in the postwar era. As its name indicated, its visual clarity of construction displayed the unity and singularity of 'whole earth' at one glance. At the same time, the model emblematized globality as a structure, as a construction made from numerous single parts whose origins and place of montage were anything but universal. Unisphere was a local model that illuminates how global structures are assembled from individual components whose origins of composition are often concealed. Unisphere, at the beginning of the 1960s the world's largest stainless steel structure, was sponsored, designed and built by the corporation United States Steel (USS). Unisphere's location, New York City, and the context of its creation, the World's Fair of 1964–5, demonstrate that the model presented a specific historical and geographical version of universality.

With regard to its location, Unisphere may also be understood as an allegory of the transition from the Eurocentric worldview of the Imperial Age of the nineteenth century to a world centring on an American perspective during the course of the twentieth century. While until the First World War universality had mostly been synonymous with a European view, universalistic ideas in the interwar period and in the Second World War, and to some extent also in the Cold War, would be equated with an American view of the world, which developed into the general view of the wealthy industrial countries in the Western and the Northern Hemispheres and assured that the 'only conceivable future of the world' seemed 'to consist in its progressive Westernization'.[18] The expansion of Western development concepts over the world and the political practices of 'technology transfer' from the centres into the peripheries, into the so-called underdeveloped countries, renewed former colonial relationships.[19] The US could claim to be universal in the assertion of attending to a common good, while striving for world leadership.

The affable unifying expressions of 'universality' and 'globality' are highly suspicious with regard to the intricate power relations they involve and the political tensions they habitually conceal. Processes of globalization need to be addressed as messy bricolage saturated with power. Again, Unisphere forms the perfect analogy. The model of a unified earth was composed of more than 500 structural elements, assembled into meridians and then covered by steel plates representing the continents. The capital cities of the most prominent nations of the world were illuminated. 470 tons of steel were used to build this construction. Unisphere reached a height of 140 feet, corresponding to twelve stories, and a diameter of 120 feet. The model represented the new and strong alliance of science and technology, industry and national government in the second half of the twentieth century, which has been termed 'big science'.[20] Unisphere paradigmatically demonstrates how science and technology increasingly engaged in the global transformation and configuration of the world and the earth after the Second World War. It also stands for the ways in which inequalities were preferably dealt with at the time: Since the Continental planes and rock masses were not evenly distributed across the terrestrial globe, enormous tractive and compressive forces developed on the structure during its creation. These global tensions were resolved numerically by using the latest high-speed computer technology. Computers processed the large numbers of mathematical equations that had to be simultaneously solved to control the design. A single one of these mathematical problems required the use of 670 equations processed simultaneously. United States Steel stated:

> From beginning to end, Unisphere demanded entirely new techniques to solve entirely new problems. At no point could U.S. Steel engineers go to the book for their answers. There wasn't any book. But when the time came to put the pieces together, they fit. They fit each other, they fit the theme of the New York World's Fair, and they fit the modern notion that no structural design problem is too tough to solve, given the right technical know-how, and the right facilities, and the right steels.[21]

In the 1960s, globality, the perception of the structure of world-encompassing issues and problems and their comprehensive solutions, was mainly conceived of as a scientific and an engineering challenge. Robert Moses, New York City Park Commissioner and chief architect of the New York Fair, referred to the musical version of Jules Verne's novel *Around the World in Eighty Days*, shown in the summer programme of the fair, as a romantic reminder, underlining 'that it was only 90 years ago Phileas Fogg captured the popular fancy with a bet that he and his valet, Passepartout, could circle the globe in 80 days by boldly and ingeniously catching every available form of transportation – and won'.[22] The progress of humankind since the 1870s could, according to Moses, be assessed by taking in the acceleration in orbiting the globe: 'The big commercial jets now make the round trip in eight days and the astronauts in 80 minutes'.[23]

A Time of Revolutions: The Environmental Turn

The astronauts of the US Apollo missions of the 1960s and early 1970s, revolving around the earth within mere hours, were the first to take photographs of earth as a whole. A photograph of 'Earth Rising', taken during the flight of Apollo 8 from the dark side of the moon in 1968, supplemented satellite exposures from the year 1967. The manned landing on the moon by Apollo 11 in 1969 transmitted images of the earth live to the world. Finally, the famous image of 'Full Earth' as first seen by the crew of Apollo 17 spread across the world in 1972. 'Blue Planet' turned into an icon of the earth's new exposure to risk and loss in the second half of the twentieth century.[24] The pictures displayed the earth in the dynamic green and blue colours attributed to three billion years of transforming sunlight into the processes of life. James Lovelock would later state that it 'was this kind of evidence from space research that led me to postulate the Gaia hypothesis',[25] his holistic view of the earth as one single organic whole. It is not without irony that it was this distant vision of the earth that focused attention on earthlife and its fundamental conditions, believed to be unique within the universe.

In environmental studies scholarship, this change of perspective is sometimes called the 'Second Copernican Revolution'.[26] With regard to the Copernican Revolution of the sixteenth century, having displaced the earth from its centre in the solar system to a more peripheral existence among a number of other, similar planets, this second revolution is said to have brought the earth back to the core of human attention. Both 'revolutions' overthrew the prevalent views of their times. While the first weakened the supremacy of the earth and of humankind in a larger cosmology, the second revolution, in a holistic sweep, was again partial to earthly life and living conditions – it brought the earth back into the centre of the human universe. Unisphere is but one example of the new meaning of 'whole earth' and the ways in which the new view on the earth became manifest in the 1960s.

The 1960s and 1970s have been framed as a time of revolution in several regards. In the US, such protest movements as the black civil rights movement, the women's movement, the youth and students' movements or the New Left, have been described as revolutionary transformations.[27] Recently, scholars of contemporary history have begun to address the 'the long seventies' as a 'time of transition'. They emphasize the fundamental structural societal changes at the end of the 'Golden Age', which characterized the classical period of high industrialization in the immediate post-war time from the mid-1950s to the mid-1970s. Some historians have identified major ruptures not only in 1968, but also in 1974, when the first dramatic economic crisis of the after-war period led to a move to a post-industrial, service-based economy.[28] Above and beyond, the 1970s are distinguished by the end of the Vietnam War, by the onset of a politics of *détente*, and by a gradual decline of the strictly bipolar world order – characteristics which support Maier's analysis of the time as forming the end of 'territoriality' as a structural concept.

Environmental historians have conceived of the time around 1970 as the 'threshold' period of environmentalism, stressing the sense of urgency that resonated in the demands of contemporaries for immediate action to secure the planet's survival.[29] The late 1960s and early 1970s saw the emergence of an environmental movement in the Western world that called for an immediate and radical assessment and immediate cutback of the hitherto industrial and economical mode of the so-called developed world.[30] Environmental activists and scientists alike addressed air and water pollution, nuclear power and atomic waste, chemical poisoning, heavy fertilization and the use of pesticides; they placed topics such as resource depletion, the overuse of fossil fuels, the overfishing of the oceans, the loss of species diversity, the 'greenhouse effect' and the melting of polar ice-caps on the international agenda of industrial nations. Questioning post-war economic growth and a technological development that had relied extensively on science-based innovations and had promoted a rather unproblematic view on science and technology, the newly forming environmental discourse was highly politicized and also extremely polarized to the origins and the possible solutions of perceived problems. Alarmist, in parts apocalyptic proclamations of crisis by the 'Survivalists' of the so-called 'Counterculture'[31] opposed the technocratic optimism of the 'Prometheans' or space age 'Cargoists'.[32]

The 'years of decision' around 1970 have been portrayed as a historical turning point, an 'environmental revolution'.[33] Depending on geographical location and problem perception, the '1970s diagnosis'[34] highlights certain years and events as significant. Nineteen seventy-two was the year in which the UN Conference on the Human Environment in Stockholm took place and in which upsetting studies such as *The Limits to Growth* and the *Blueprint for Survival* were published. Nineteen seventy-four was claimed as the first UN World Population Year; it also marked the beginning of the demographic transition in the Northwestern world. The years 1973 and 1974 are often underlined for the oil price shock, which is said to have put an end to the 'age of ecological innocence'.[35] Pressures were perceived as imminent, and future effects were held to be long and lasting. Decisions made at this point, according to the general opinion of the time, would determine the fate of the earth and of human life on the planet for decades and even centuries to come.

An obsession with the future and with future management took hold, mirrored in the new field of future studies or 'futurology'.[36] What would eventually be subsumed under 'New Age' Environmentalism was not as remote as it may seem from the hopes for deliberate planning and intervention into the planet's fate.[37] While environmentalists like Lovelock regarded the earth as a living organism and promoted the teleology of a planetary homeostasis, proponents of the latter route founded their position on traditional ideas of modernization and growth, and on a strong belief in political and technological regimes of control. Both sides relied on bold long-term prophecies and predictions on resource supplies, environmental contamination and population development.

Environmental historians have suggested the term 'Environmental Age' to express the view of the 1960s and 1970s as the rise and peak – and eventual decline – of environmental concerns and warnings about the future of the earth.[38] I use the term 'Environmental Age' not only to give meaning to a key period of the twentieth century that provides the timeframe for the account presented in this work but also to capture the diverging trajectories, the options and coping strategies as well as the fundamental controversies about what constituted an environment and an environmental problem in the first place. When the environment became a discursive object in the decades after the Second World War, also spaces and places became 'environments', which crucially differed from the 'nature' encountered in earlier nature conservation movements.[39]

Ecological or Epistemological Crisis? Challenges from Science and Technology Studies

Science and technology studies have targeted the troublesome nature of the 'nature' that the natural sciences and technologies create or affect. Science studies scholars tend to be particularly wary of the ways nature as a historically shaped entity tends to be treated as a given and self-evident category. Their work demonstrates that concepts and images of nature depend on the specific historical situations and locations in which they are addressed and developed. Nature has never been a pure material foundation to culture but has been also the effect of cultural constructions of meaning. On a similar basis, the 'environment' has been questioned as an unproblematic common ground; following the environmental historian William Cronon, both nature and environment are literally 'uncommon ground'.[40] Cronon's seminal work on the construction of 'wilderness' as one of the fundamental and unquestioned categories of North American historiography constituted a shift in environmental history writing; environmental historians argued for a diversification of environments, agreeing that there is no 'original nature' to fall back on.[41] Although a universalistic historiography of 'the environment' – in the singular – still dominates public perception, recent work has shown that environments are historically variable and are constituted in the interplay of human and material actors. Like nature, environment is the object and effect of specific architectures of power and governance.[42]

This book makes an effort to look into the uncommon ground of the environment in the Environmental Age. In so doing it will follow science studies scholar Donna Haraway's reminder that to 'be "made" is not to be "made up"' and that constructions are 'about contingency and specificity but not epistemological relativism'.[43] It is important to note that the environment envisioned by the figure of Spaceship Earth was not simply a discursive invention that could be altered at will, but rather a material-semiotic intervention, an unprecedented realization. Relics of the traditional pristine nature to be preserved survived into the times of the environment of the 'spaceship' that this book explores. In the late nineteenth century, nature 'sanctuaries' and 'reserves' defended a nature set aside from humans and for humans, and at the same time protected and patronized under human stewardship.[44] The conservative attitude of nurture and protection towards a nature wild, pure, pastoral, sublime and even divine, kept within distinct boundaries, followed national, territorial and often utilitarian ideals. A hundred years later, nature was conceived of not in immediate national and recreational terms but in the functional terms of a global ecosystem from which humankind profited in the mediated ways of systems dynamics. The 'biosphere reserves' of the early 1970s emerged out of global initiative; they were based on international networks and on different epistemological frameworks: although as exclusionary as a park or reserve, they expressed a new form of nature preservation on a scientific basis, in which selected areas were protected according to their representative ecosystem functions.[45]

In 1968 Paul Ehrlich explicitly avoided associating himself with the 'conservationists', the nature saviours who would cling to redwood trees to protect them from the mills of the big lumber companies.[46] The reason he offered was pragmatic: 'the conservation battle is presently being lost'.[47] It was of no use to shed tears on a species on the verge of going extinct. 'Americans clearly don't give a damn'.[48] Ehrlich did not expect his fellow citizens to stand up for species diversity out of mercy or for beauty; in a cultural setting dominated primarily by economic principles, environmental advocacy relied on the unambiguous indication of its utility, function or benefit. The pragmatism he exhibits is exemplary of a tendency many scientists of the late 1960s and early 1970s displayed, taking efforts in pointing out and predicting environmental deterioration, albeit cynically aware of their limited abilities to effect radical changes in the lifestyles of their contemporaries. The emerging global environment became a practical problem of national and international politics, an arena of diverse interests and conflicting priorities, and a number of sober action plans that suited various stances.

The Stockholm conference in 1972 outlined an environment that was no longer a pristine nature to be conserved; rather, the environment was a juridical and scientific object to be internationally negotiated, administrated and allocated. This new environment was tailored for research and technologically monitored and managed; it was deeply political in its constitution and closely connected to questions of development. The peculiar term 'human environment' that set the agenda for the conference illustrates how nature as detached and apolitical dissolved in the global political field of environment and development. On the one hand, the human environment portrayed humans not as masters over nature but as an intricate part of their surroundings, affected by every one of their acts on nature. On the other hand, the term exhibited a distinctive anthropocentrism; human environment took into account a nature that served as milieu to humans, a human habitat; ultimately, it considered environment as controlled or created at will by human action.

Both in the Eastern and in the Western world a greater part of the environment was regarded as nature to be 'taylored'. Reaching back to early twentieth-century ideas and rationales of scientific management, Taylorist (rationalized) and Fordist (industrialized) environments emerged that were administered according to the principles of efficient (human) organization. Environments had been 'conditioned' on smaller scales. US engineers had experimented with artificial indoor climates in the 1920s and 1930s and with closed laboratory environments in war and post-war times.[49] The new large-scale planning approaches to nature merged older ideas of social hygiene and public health with new forms of environmental and economic health. Instead of 'nature' in the traditional conservationist sense, these environments were 'second natures' or 'technonatures'; they constituted artificial surroundings considered more functional, economic, efficient and thus more perfect than 'nature' out there – and often more sanitary and clean.[50]

But environments not only responded to, they also acted upon humans. Environmental activists have been particularly susceptible to this position, arguing against destructive modes of human intervention and for a restorative approach to wilderness. Notably, scholars have long argued for a shift in perspective to nature's materiality and memory, and many examples can be found that foil the notion of human superiority over nature. The cases of nuclear weapons testing and of biological and radiological warfare during the Cold War era have given ample insight into the uncanny forces unleashed in a nature that resists human desire, changes unexpectedly and constantly challenges human ideas of stewardship or mastery. An environment bestowed with the power to contaminate human life for years to come can hardly be described as merely exposed to human will and skill; rather, it might be more suitable to say that in the twentieth century, new and unprecedented environments exposed humans to new risks and environmental hazards.[51]

These environments were more or less thriving realizations of the Environmental Age, redistributing power between human and natural actors. An environmental historiography accepting 'Taylorist' and 'Fordist' environments as mere projections of human engineering expertise tends to reaffirm the notion of powerful technological advance. In turn, a historiography that takes at face value the idea of nature defiantly fighting back is prone to turn nature into an ally by speaking in its name – if not taking the position of a military strategist viewing nature as a hostile challenge or target. A nature given power in this way will always be fought or exploited to settle human conflicts. The active environment is conceptually not far away from the sublime, alien and inhumane nature, as the animated nature is closely related to the nature that humans war against. All of these natures and environments are deeply anthropocentric, and epistemologically they would be hard to conceive of differently.

The question remains how environmental historiography can proceed without taking sides in the ongoing conflict of contradicting environments and their protagonists. Sociologists of science, at the forefront Bruno Latour, have provocatively and programmatically demanded 'the end of nature in politics' and thus asked to set up an entirely new 'constitution' to conceptualize the interaction of human and non-human actors.[52] Following Latour – that is, seriously challenging the 'modern constitution', the two separate houses of 'nature' and 'society' that constitue the modern world – requires one to take 'nature' (in the singular case) out of the hands of 'science' (in the singular case), since a nature subordinating culture and politics – the facts first, the fictions, values and intentions second – suspends a truly political discussion.[53] Latour's audacious suggestion comes as one more 'revolution' in the face of one more crisis: the epistemological crisis. Latour maintains that the ecological crisis did not manifest itself as a crisis of nature or environment, but rather as a 'crisis of objectivity',[54] as a crisis of the irrefutable, the undeniable, the unquestionable – the natural. Environmental crises and environmental catastrophes point us to the fissures and shortcomings of the modern world and its technoscience.[55] The gaps and inadequacies of processing and possessing the modern objects are not reflected, and consequently they erupt as political conflicts about the societal relations towards what is considered the environment.

Latour's approach is appealing in that it offers to reformulate the question about a real and original nature and environment as a question about the processes of modern scientific and technological constructions of evidence and certainty. Addressing the problem of legitimacy of the modern sciences and of modernity does not end in postmodern arbitrariness and relativism, but yields very productive effects. The humanities and cultural studies are more than ever called upon – and are able – to contest the monopoly on nature that the natural and engineering sciences have successfully claimed for so long. The humanities are capable of questioning how technoscientific knowledge and practices are constructed to appear universal and true; they inquire how modern technosciences authorize their assertion to speak in the name of a nature, which in the same breath they organize and adjust. Moreover, the humanities explore how the environment and environmental problems can be conceived of as historically and culturally constructed and yet provoke real material consequences under real material conditions. Let Haraway remind us again that constructions are about contingency and specificity, not about epistemological relativism. This perspective, she observes, is conditional for the strong political claim that not 'all views and knowledges are somehow "equal", but quite the opposite'.[56] Some environmental histories can be more problematic in their consequences than others.

Spaceship Earth: Environmental History Beyond Sufficiency and Efficiency

In this book I explore the discursive power of Spaceship Earth in the Environmental Age of the 1960s and 1970s and until the early 1990s. Adopting perspectives of historical science and technology studies, I raise the question of how the figure of Spaceship Earth framed and structured the terms, the possibilities and the conditions of environmental discourse. Through this approach I mean to contribute to an environmental historiography that avoids the fallacy of simply taking sides in the ongoing debate between Survivalists and Cargoists, between environmentalists and technological optimists, between supporters of sufficiency and advocates of efficiency. Neither can it be my aim to resolve these debates. I would like to tackle a more fundamental problem by questioning the prevailing idea that earthly confinement can be confronted technologically and solved by creating controlled environments as living spaces. Spaceship Earth is the perfect expression of the observation that a nature violent and sublime and an environment conditioned by engineering skills are two faces of the same modernist coin. Outlining the narrative qualities of Spaceship Earth as a myth will serve me to point out the double strategies involved in framing the earth as a spaceship: Spaceship Earth presented the planet not only as a singular site in the universe, but it also signified a growing scientific fascination with natural-technological environments on earth and beyond.

Spaceship Earth became the core of a mythology. Myths are not fictitious stories, and they do not stand in opposition to reality. Myths are culturally shared narratives that describe collective human realities meaningfully. Their messages may be equivocal. This book claims that in a time concerned with limits as much as with opportunities Spaceship Earth became popular precisely because the figure drew and profited from antagonist positions and synthesized different standpoints: Spaceship Earth represented crisis and progress at once, enabling environmentalists and technocrats, doomsayers and technological optimists to argue seemingly opposite positions in the same terms. Spaceship Earth was not simply a metaphor of perceived environmental problems but also presented possible environmental solutions by opening up new discursive trajectories for future living spaces. Spaceship Earth imagined a global, sustainable, natural-scientific environment, closed and controlled like a space capsule.

In the high times of the Space Age the spaceship seemed appropriate to combine the notion of life's fragility on the one hand and of the triumph of science and technology on the other. Inside this discursive frame, it took only a small step to imagine the newly discovered planet earth as a spaceship. The US ambassador to the United Nations, Adlai E. Stevenson, used the image of the spaceship in his last speech before the Economic and Social Council of the UN in Geneva in 1965. In an appeal to the international community Stevenson referred to the earth as a little spaceship on which humankind travelled together as passengers, dependent on its vulnerable supplies of air and soil.[57] In 1966, the English economist and political scientist Barbara Ward chose the term to make a plea for a new 'balance' of power between the continents, of wealth between North and South, and of understanding and tolerance in a world of economic interdependence and potential nuclear destruction.[58]

The true architects of Spaceship Earth, however, used the term not as a metaphor of vulnerability and community but to describe an innovative technological *model* of a natural environment yet to come. The architect and designer Richard Buckminster Fuller allegedly used the term already in 1951 in a discussion about the US space rocket programme and presented a paper on 'Spaceship Earth' in 1963. In 1969 Fuller published his *Operating Manual for Spaceship Earth*, summoning the engineering elite to take control of an environment in bad repair.[59] The economist Kenneth E. Boulding in a programmatic lecture on 'The Economics of the Coming Spaceship Earth' held in the year 1966 dated the 'global nature of the planet' back to the post-war times and to space flight.[60] At present, he stated, mankind experienced a 'transition from the open to the closed earth'. Boulding chose the spaceship as the central image to promote the 'closed earth of the future', suggesting to foreclose the wasteful 'cowboy economy' of the past for a frugal 'spaceman economy'.[61]

Boulding's vision contributed in a specific way to the discourse of the fragility and limitedness of planet earth on the one hand and of its future management and operation on the other. The self-sufficient spaceship became his central model to denote the economy of the future earth. At the same time, his suggestions speak of his strong belief in systems ecology, in the prevalent sciences of his time and in their power of efficient planetary regulation. His image of the earth as a single global system, spatially and functionally closed, combined ideas developed in ecosystems research, in economics and in cybernetics since the Second World War. On the physical basis of biological processes an economic-ecological reasoning in terms of a circulatory system or earthly metabolism became conceivable and calculable. This system of stocks and flows of energy, information and matter contained older ecological concepts of a holistic nature, but developed them into the environment as a self-adjusting machine.

Spaceship Earth offered the blueprint for an economy of 'circulation' and for a technology of flows, of material exchange and renewal, for the earth's living space. A hybrid of Gaian organism and planetary technoscience, Spaceship Earth needs to be seen as a veritable 'cyborg' metaphor, a figure that symbolically and materially dissolved political and social contradictions and conflicts.[62] For the same reason I would like to hypothesize that Spaceship Earth disintegrated the Latourian nature–culture divide. On all accounts I put forward that Spaceship Earth presented an appealing possibility of conceptualizing a historically specific version of the organic and the technological as a hybrid. The figure of Spaceship Earth linked the concerns about the planet and about earthly confinement with the imaginations and technical solutions of planetary maintenance and control.

A second central feature of Spaceship Earth as a cyborg figure concerned a degree of creative self-organization and optimization based on the increasing autonomy of technological potential directed at the earthly environment. The spaceship environment was a highly contested space, scientifically experimentalized, modelled and monitored; it was engineered, economized and politicized; and it was enrolled into military considerations. Its organic elements became environmental sites, unsettled, wasteful or disposed of. Spaceship Earth formulated prescriptions as to the passengers it would carry. Paul Ehrlich and Richard Harriman organized their entire book *How to be a Survivor: A Plan to Save Spaceship Earth* in 1971 around the metaphor of spaceship earth, from the 'Size of the Crew' to a new culture of 'Spacemen' needed.[63] The constitution of the environment as a coherent and potentially controllable global system enabled the distinction between the necessary and the disposable, between nature and its equally functional and possibly more potent and innovative technoscientific substitutions.

Circulation – Storage – Classification: In Pursuit of Ecological Blueprints

Spaceship Earth indicated a fundamental change in the perception of the earthly environment as finite, while serving as a model of a specific scientific-technological management of life and living space in its allegedly natural limits. From the newly generated knowledge about the finite dimensions of planet earth emerged new regimes of efficiency. I explore technicity and functionality, site and selectivity in the cultural history of the ship and the spaceship. My focus is the economic accounting and balancing of nature, of life, and of environment in terms of spatial confines and resource scarcity in the sciences of ecology and human ecology. Following Foucault's concept of biopolitics, I analyse the rationalizations and the operations in which debates about population growth and resource depletion intersected with ecosystems management and space technology and condensed into new technoscientific simulations and substitutions of human living space. This work carries forward the current debates about a New Global History, albeit from a distinctive perspective, expanding the history of 'globalization' in the twentieth century to historical science and technology studies and to the cultural history of the environment in the second half of the twentieth century. Taking this perspective, I aim at contributing to an understanding of environments as technoscientifical constructions. I also seek to make a contribution to a contemporary historiography which has settled on caesuras around 1970 and 1989/90 but is only just beginning to take the tremendous and unprecedented scientific and technological potentials in this period seriously.

Proclaiming the 'epoch of space' in 1967, Foucault seemed to have apprehended the historical focal shift towards the world as a community sharing a common destiny on an absolutely limited planet. He introduced the example of demography to argue that 'living space' – he also called it the 'human site' – in the twentieth century was not so much a question of fretting over the particulate securing of territorial property as a question of a new rationality of global biopolitical *siting* and *placement*:

> This problem of the human site or living space is not simply that of knowing whether there will be enough space for men in the world – a problem that is certainly quite important – but also that of knowing what relations of propinquity, what type of storage, circulation, marking, and classification of human elements should be adopted in a given situation in order to achieve a given end.[64]

As a corollary of the problem of 'living space' Foucault anticipated the discipline of 'human topography' rising to account for the siting and placement of 'human elements' according to a combination of mathematical, geographic, technoscientific and economic expertise.[65] I set up my story of Spaceship Earth according to the principles and practices of *circulation*, *storage* and *classification* that Foucault anticipated, and according to the modern disciplinary expertise he associated with these principles. Spaceship Earth reformulated the theme of contested living space along the lines and divides of biological eligibility commensurate to the notion of limited global ecological 'capacity'. I will associate Foucault's perspective with specific cultural-technoscientific practices that created a new architecture for conceiving of the environment as a global *interior* space: 'Circulation' I will relate to the physical, technological and economical concepts in ecology that transformed the earth into a closed metabolism and the earth's biosphere into a self-sustaining 'life support system'. 'Storage' I will relate to the geography and the mathematics applied in solving the global equations and ecological balances of human ecology facing the earth's limited 'carrying capacity'. 'Classification' I will associate with the biotechnical selections and exclusions enforced and implemented by those spaceships that sustained life support systems elsewhere.

Closely adhering to the Foucauldian proposal I concentrate on scientific practices and related technologies engaged in forms of ecological economics: systems ecology, human ecology and biospherics were concerned with accounting for natural earthly limits. I narrow my story to an exploration of the setting in the United States, as the plot relies on American academic debates within the natural sciences and on US space politics. Not only did Spaceship Earth emerge in US environmental discourse, but the United States also became the discursive axis of Western environmentalism and environmental sciences. From US discourse Spaceship Earth propagated and became a familiar term in different Western languages.[66] I will examine material from the fields of ecology, economics, biology, physics and cybernetics between the 1960s and the 1980s and ask how within these fields the thematic triad of the era – 'Population, Resources, Environment'[67] – was formed, and how it interconnected with larger cultural and political themes of living space and space flight.

Unquestionably this approach will only represent a small section of the era's environmentalist notions and politics, namely the segment of environmental discourse that was concerned with ecological problems about absolute global limits and expecting models and blueprints from science and technology. This is not to say that no other dominant positions existed at the time in question, positions which perhaps even claimed to offer strong culturalist alternatives to the very scientist accounts of earthly problems and solutions I present here. Still, I hold that the figure of Spaceship Earth and its atmosphere of vulnerability, technology, hierarchy and temporality structured scientific disciplines and societal debates and assigned credibility to the interventionist claims of science and technology in environmental discourse. At a time when the exuberant and procreative qualities of nature were in question Spaceship Earth and its concentration on 'the planetary' revised – and perhaps replaced – much older images of a fertile and nurturing 'Mother Earth'.[68]

Adopting an environmental history perspective and a science and technology studies approach I pay attention to the scientific practices and techniques of calculation and visualization that produced self-evident environments like that of the spaceship. As a cultural historian, I am interested in the conviction established by scientific representations, to the ways of communication between the sciences and their broader cultural contexts, and to the cultural meanings implied in the new symbolic organization and order of nature. Next to the ecological sciences I draw on material from contemporary culture and politics that sheds light on the possible or desirable futures of the earth. I incorporate different images of the earth, contemporary projects on the architecture of earthly living space design, fictional literature and cartoons. I also discuss a number of

science fiction works throughout the book that reflect on the utopian and dystopian aspects of the anticipated 'ecocide'. I will not be concerned with general
space science fiction focusing on life in space and only alluding to a far-away
earth. The fictional works I study are mainly set on the earth; they allow me
to identify some of the persistent themes of ecological discourse – growth and
regression, capacity and balance, containment and conservation, selection and
exclusion, extinction and survival, to name but a few – and to explore how these
themes were culturally contested or corroborated in filmic stories.

Spaceship Earth was never depicted in one single image. Spaceship Earth was
presented textually and visually in very different ways. Among its expressions
was the global and universalistic idea and model of the Unisphere as well as the
picture of the vulnerable Blue Marble. It will be part of the cross-disciplinary
endeavour of this book to sort through some of the textual and visual ideas and
images that contributed to the success of the spaceship and the Spaceship Earth
as a *disposition* of its time. The interdisciplinary effort deliberately cuts across the
fields of the humanities on the one hand and the natural sciences and technologies on the other. In addition, the material presented will not be unpacked in a
strong chronological order; rather, it will be arranged synchronically with the
idea to emphasize the similarities and equivalences in adjoining discourses that
forged meaningful environmental realities.

Temporally, the discourse of Spaceship Earth spans a short time period of
thirty years between 1960 and 1990. When the Environmental Age came to its
end, Spaceship Earth was abandoned. The figure more or less disappeared from
popular science and culture, surviving only in very special scientific discourses
of the geoclimate and geospace sciences. Spaceship Earth was discarded with
the popularization of the concept of 'sustainability' in the mid-1980s. The new
programmatic vision strongly emphasized political participation, locality and
diversity over ideas of a universal planetary management. Nevertheless, the concept of sustainability, strongly accenting political participation and economic
development and growth, continued much of the discourse of environmental
sustenance and technical maintenance that had been at the heart of Spaceship
Earth. The 1990s witnessed the end of the Cold War and the rise of a discourse
of globalization that depicted a shrinking globe as a result of growing intricate
networks of communication and financial transactions within a promising New
Economy. The new visions of a 'sustainable' development were grafted onto an
ecological reality that was firmly based on economical reasoning.

The Plot: Spheres, Systems and Reserves

The ship is the recurring figure underlying the story of Spaceship Earth. To argue that the plots of ecosystems science, human ecology and biosphere technology all centred on motifs of the ship – the ark, the lifeboat, the spaceship – I will begin my account by highlighting the significance of the ship in Western culture to explore the ways in which protection and exposure merged in the motifs of the ship and the ship's voyage. Under the title 'Containment', I argue in Chapter 2 that the figure of the ship linked narratives of exposure, fragility and transience on the one hand and expedition, exploration and expansion on the other, and that these narratives were continued in stories of the spaceship, which then became meaningful in environmental discourse. Spaceship Earth, so my argument runs, continued and expanded the figure of the biblical ark. At a moment when the earth was discovered as a paradise soon to be lost, the spaceship expressed the vision of a modern-day ark to preserve and extend earthly life by new means.

Paradoxically, the ship's containment also featured as the crucial precondition of expansionist moves. Spaceship Earth marked the earth as a temporary environment and opened up the horizon for the survival of mankind elsewhere. The chapter draws on the mythology of the 'frontier' that is enclosed and repeated in many ship and spaceship narratives. The frontier mythology suggests that the heart and fate of a growing nation is to be found at its edges, that a people is defined by the territory it must conquer and cultivate. The history of the American Frontier can be understood as a narrative of national origin that was crafted to explain why the people of the United States were destined to colonize a newly discovered continent. I am interested in how the concept of manifest destiny, of a fate that fulfils an ordained purpose, is mirrored in the figures of the spaceship and of Spaceship Earth. These considerations can only be outlined roughly in the following chapters, as they point towards more complicated questions that this book cannot explore in detail. These questions involve the problem of whether and to what extent the cultural imagery and the mythology of space flight in the twentieth century derives from, reflects or embodies the Judeo-Christian religious tradition that shaped Western culture, and whether and to what extent space flight is characterized by a teleological understanding of development and expansion, which for the earth and for humanity promises either salvation or doom.

Under the title 'Circulation', Chapter 3 discusses the figure of Spaceship Earth as a model of cybernetic regulation and control of nature and life in the emerging ecosystems sciences. Systems ecology turned the earth inside out to reveal its metabolic principles and material flows. I will discuss how a concept of the biosphere as a closed interior space could emerge, a complex self-contained metabolic sphere that framed the earth's environment as a 'life support system' in analogy to spaceflight. The chapter opens with the work of Eduard Sueß who first introduced the biosphere as containing the whole of the living world around the turn of the twentieth century. It briefly sketches how Vladimir Vernadsky turned this sphere into a system in the 1920s. The focus of the chapter lies on the rediscovery of the biosphere in the early 1970s and on the question how this new environmental concept balanced a sober eco-systems calculus against conservational plans and programmes. The notion of a global environmental system – of a defined, efficient and sustainable metabolism of energy, information and material flows – supplied a scientific basis for creating inventories and rational schemes of optimal use of biosphere resources. The chapter draws on the example of the new equivalences of economical, ecological and technological environments with international programmes to study and preserve the biosphere as a global natural resource in preparation of and subsequent to the Stockholm conference in 1972. The 'Man and the Biosphere' programme designed 'biosphere reserves' in a coordinated world network representing ecological functions in the ecosystem of planet earth. I argue that the biosphere reserve not only set aside especially fragile and threatened parts of nature by demarcating restricted and protected areas but also reconciled nature conservation with the utilization and enhancement of nature.

The following two chapters will investigate how an economy of obsolete and useful single parts of the environmental metabolism gained ground in the discourse of the limited ecological 'carrying capacity' of the earth. Under the header of 'Storage' Chapter 4 inquires how by applying metaphors of the ship and the lifeboat 'overpopulation' became an evident and calculable problem. I will address the conditions and implications of aggregating humans into 'populations' through statistical means. Engaging with statistical constructions of world population and population growth in the twentieth century, I discuss the 'biological law of population growth', which corroborated predictions of a 'population explosion' and demands for 'population control'. The biostatistical model and curve were developed in experimental animal population biology to describe the self-limiting growth of self-contained populations over time and informed the human development studies of the 1960s and 1970s. The chapter explores the economies inherent in the 'law of growth', arguing that the law structured and insinuated a biopolitical system of classification and allocation of human lives according to a 'lifeboat ethics'. I analyse the scientific strategies of abstraction, reduction, formalization and visualization effective in the growth law and trace its increasing power to account for growth phenomena in general. Through these scientific methods population issues were assigned to the realm of the life sciences, to the effect that sociopolitical approaches were overruled by bioeconomical ones: statistical accountability constructed populations as assessable and either valuable or dispensable on a global scale.

Chapter 5, titled 'Classification', explores the project of Biosphere 2 that was launched in the Arizona desert in 1983 with the aim of constructing a self-contained and self-sustained ecosystem modelled on the earthly biosphere. Biosphere 2 was an ecological experiment of a new type and scale: On a site of three acres, seven defined earth ecosystems or biomes were established and sealed under a glass dome. Based on an invisible technological infrastructure the living laboratory was populated by nearly 4,000 animal and plant species as well as eight humans in order to understand and, ultimately, to operate ecosystem processes on planet earth and beyond. The human-natural society was designed as a prototype of a space colony that would eventually enable its deteriorating predecessor Biosphere 1 to create robust offspring. Evolving in a Darwinian fashion, future biospheres would allow small human settler societies to migrate to other planets.

Biosphere 2 merged the earthly biosphere with a veritable spaceship environment in the attempt to construct a high-tech surrogate earth operable at any terrestrial or extraterrestrial location, a truly luxurious artificial space in the sense of the Renault Espace: sheltered, gleaming, cosy, air-conditioned and fully controllable. Biosphere 2 not only acted as a storage space where natural and human elements and technoscientific knowledge were assembled and combined for the replication and substitution of earthly nature. Biosphere 2 also implemented new conditions of admission and entitlement to life within the enclosed miniature earth. The model of sustainability employed in the second biosphere emphasized not completeness but (systemic) integrity. Access to the biosphere, once defined as the sphere that contained all life on earth, was subject to strict processes of selection to determine the most useful, collaborative species and to recruit a species' best-designed representatives. The chapter will discuss how a life support system based on ecological and technological efficiency reshaped demands on the natural and social environments inside and outside of Biosphere 2's glass dome.

The privately financed project of Biosphere 2 has often been perceived as the eccentric endeavour of a handful of environmentalists. Notably, however, the project also involved some of the pioneering systems ecologists and ecological architects of its time. Patents were obtained and papers were published in peer-reviewed journals. Although it aspired to put into practice a radical alternative to contemporary environmentalisms, the project was not opposed to but quite in line with ecosystems and cybernetic ideology, relying profoundly on material sciences, on engineering expertise and on the latest technologies. Biosphere 2 illustrates a paradoxical shift in the meaning of the 'biosphere reserve' brought about by ideas of nature protection and perfection. The reserve not only refers to nature set aside but also to the exclusiveness of the 'reservation', the claim, alleged right or title to make use of nature and environment under emergency rule. Biosphere 2, conceived of as a modern-day ark, sheds light on some of the intricate and sometimes subtle relations of ecology and spaceflight in the era of environmentalism as well as on the shifting relations of 'first' (pure) and 'second' (technologically reproduced and substituted) natures in this historical process and the power relations involved in the interplay of modern natures and technologies.

The sixth and final chapter, titled 'Departure', contends that the obsessions and visions about the future held in the realms of the Space Age and the Environmental Age ultimately collapsed in strategies of leaving the earth – albeit, perhaps, not in the spaceships that were designed for this purpose. Since the 1990s a variety of novel ideas and images has taken the place of Spaceship Earth. Although the problem of global space has not eased but rather increased in the past two decades, Spaceship Earth and the idealized habitats it offered have turned out unable to carry the weight of the world and its present problems. Today's conditioned spaces no longer claim to contain and control, much less move the entire planet. The new sites are highly individualized and exclusive. They form withdrawn and insulated sanctuaries amid increasingly wasted earthly environments.

2 CONTAINMENT: THE SHIP AS A FIGURE OF ENCLOSURE AND EXPANSION

A Conservationist Mission

The opening sequence of a science fiction film released in 1972 begins with close-up views of an Edenic garden of plants – plentiful, precious, pure and peaceful. It soon becomes clear, however, that the richness on display is as unique as the biblical paradise. As the camera draws back, we see that this botanical abundance is quite limited, contained within a large glass dome. Pulling back still further, the camera reveals that the encapsulated environment is actually situated in deep space; the dome is part of the *Valley Forge*, a huge American Airlines space freighter.

Accompanied by majestic music, this sequence culminates in the revelation of the awe-inspiring extent and importance of this US mission. A solemn voice reads out a declaration written at the beginning of the twenty-first century, in which the last surviving forests on earth are dedicated to a conservationist journey through outer space that, as the story begins, has been in progress for eight years. Astronaut and botanist Freeman Lowell, one of four spacemen aboard, tends the precious cargo that will one day be returned safely to its earthly home. Freeman is a true disciple of the environmental movement of the 1960s and early 1970s. He still remembers when vegetables had taste, smell and colour, when the air was fresh and the skies were blue (a time evoked by the film's soundtrack of Joan Baez performing folk music). Through Freeman we learn that, despite the warnings of environmental activists, the earth of the future has become a bleak and uniform place with a homogeneous temperature of 75 degrees Fahrenheit. The planet is so densely populated that it has grown into one massive, completely defoliated city; trees and plants are no longer essential for (human) life, as nutrients are now laboratory-manufactured.

The startling contrast between the richness of life and the fragility of its artificial environment, as well as the consequences of changed attitudes towards nature, are the issues on which the 1972 movie *Silent Running* focuses.[1] The film was produced amid heated debates on resource scarcity, environmental pollution and overpopulation. It reflects the popular images of impending ecological catastrophe and the questionable survival of humankind in the Environmental Age. It is certainly no coincidence that the film is set entirely on a spaceship, which had become a major symbol of both the fears and the hopes associated with the earth's transformation into an endangered planet. At the height of the Space Age, the spaceship merged notions about the fragility of life with the triumphs of science and technology.

The film draws both notions to their extremes and projects them into the near future of the second millennium. The spaceship, a high-tech carrier, represents the high hopes and aspirations connected to technological progress. Being the only known technical means to transgress the limitedness of a deteriorating earth, the spaceship is portrayed as following a trajectory, a path away from the earth to the stars. The journey is a mission showing direction and purpose. The spaceship takes the place of a conservatory that contains the life that was once known on earth. At the same time, the spaceship marks a provisional solution. It is at the core of an impossible voyage, rootless within plain and desolate surroundings, dark and lonely, isolated and monotonous, its future uncertain. The movie depicts the ship in all its tragic paradoxes: the spaceship represents the ultimate upshot of progress; it is entrusted with the burden of salvation of the earth from the ultimate environmental degradation, which in turn is effected by the massive technological expansion that brought about the space vehicles in which humankind distanced itself from the earth.

Centring on the tragedy of the spaceship, the movie identifies the clash of 'first' and 'second' natures to be the predicament of the Environmental Age. A nature rich but threatened meets an artificially created environment to compensate for deterioration and substitute for loss. The figure of Spaceship Earth expressed a similar predicament for planet earth. Although publicized as a metaphor to denote the fragility of the earth, it also carried the hopes of being able to manage the planet by scientific and technological means, like a spacecraft. Imagining earth as a spaceship made it possible to combine concerns about the planet with visions of global control in order to address the question of mankind's 'survival'. Denoting the singular place of the earth in the universe, Spaceship Earth also signified the growing scientific fascination with natural-technological environments on earth and beyond. Spaceship Earth marked earth as a temporary environment and introduced the idea of human survival elsewhere, a survival that would be based on rational scientific management.[2]

Central features in this plot are the image and practice of the voyage. A ship's voyage is characterized by the 'remarkable combination of security and vulnerability' that Barbara Ward pointed to in 1966 when she described the earth as a single spaceship on a precarious 'pilgrimage through infinity'.[3] In this chapter I explore some of the ways in which protection and exposure merged in the motifs of the ship and the ship's voyage. The chapter's trajectory follows the meanings of the ship in Western cultural and literary narratives as they are linked to fragility and transience on the one hand and to exploration and expansion on the other. I look into the cultural history of the ship as a vehicle of hope and rescue, flight and fate. This strand explores the issue of containment associated with the figure of the ship. A second strand follows the voyages of discovery and exploration and their relation to spatial expansion. I study the themes of boundlessness, illimitability and infinity expressed by the figure of the ship. This strand deals with the creation of a ship's environment in relation to the frontier mythology. A discussion of the mythical themes of ship narratives will illuminate the double strategies involved in the ship's voyage: the ship is at once a static, self-contained site and a means of knowing the earth; the ship encloses space, it explores the realms of the unknown and it transforms earthly space into bounded territory.

At a time when the earthly environment seemed like a paradise in jeopardy, the spaceship, like the biblical ark, held out the hope of preserving life in all its diversity and of settling it elsewhere. I am interested in the spaceship as a frontier vehicle and in how in the 1960s the spacecraft supplemented the ship and expanded traditional frontier narratives. The ship's fragility and transience, exploration and expansion established trajectories that linked the promises of modern arks with pragmatic practices of selection. I address the biopolitical and bioeconomical plots of the ship in which valuable human life was distinguished from disposable 'surplus'. These plots will be taken up in subsequent chapters of this book to explore how population ecology developed new ethics and practices of determining capacities and overloads by means of the lifeboat and the spaceship.

Narratives of the Ship in Western Culture

From the early modern voyages of discovery to the Apollo missions the ship has served as a reservoir of collective memory and imagination. Symbolizing spatial expansion and exploration of the unknown as well as fragility, transition and transience, the image of the ship has been at the heart of Western culture's most powerful narratives. Michel Foucault considered the ship to be the 'heterotopia par excellence'. *Heterotopia* was his term for an exceptional site that exists within the world and, at the same time, lies far remote from or beyond it. Heterotopias, according to Foucault, are in relation with all other places and spaces, and yet in opposition to them. The ship he described as a 'floating piece of space, a place without a place, that exists by itself, that is closed in on itself and at the same time is given over to the infinity of the sea'.[4]

Foucault might have had in mind the *Flying Dutchman*, the legendary ghost ship that was condemned forever to roam the seven seas, never to enter a port, and that brought doom to all seafarers that would come across it. He may have thought of Sebastian Brant's 1494 moral satire *The Ship of Fools*, a story about a weak and evil community that falls to the seven deadly sins and a range of other, worldlier vices, on its journey to oblivion.[5] Maybe he thought of Plato's ship of state, the metaphor of a society governed by the expert few. Whether he referred to the *Epic of Gilgamesh* and the perilous crossing of the Water of Death in search of immortality, or to the *Odyssey*, Homer's epic poem about Ulysses's tormenting journey home to Ithaca that took him to the ends of the earth and to the land of the dead: the ship, the frail vessel in the midst of stormy seas, broaches the issues of authority and control, of providence and fate – not to forget human presumptuousness and subsequent downfall. In this regard the twentieth-century example of the *Titanic* has reached mythological status; the epitome of all sunken ships throughout history conveys more than other stories of shipwreck the frailty of the boundary between the interior and exterior worlds that technological prowess will never entirely master.

The German philosopher Hans Blumenberg explored shipwreck as a metaphor of existence, placing shipwreck in the midst of a broader nautical imagery of islands and havens, storms, depths and shallows of life, cliffs and calms, navigating and capsizing, salvage and sinking.[6] Although inhabitants of the mainland, humans tend to imagine their worldly existence in terms of seafaring, despite the sea figuring as the uncanny realm at the edges of the known world that is populated by thunderous gods and mythic monsters. Blumenberg explains the paradox by exploring the imagery of seafaring as a violation of borders. In this imagery the odyssey pictures the effect of the arbitrariness of nature's forces, while shipwreck illustrates the consequences of humans daring to leave their assigned place in the world. Shipwreck is the result of a divine power punishing the blasphemy of nautical transgression or hubris.[7] Following Blumenberg, the nautical imagery took an essentialist turn through Friedrich Nietzsche's view that humankind had embarked, burned its bridges and left the land irreversibly. 'The metaphor of embarkment involves the suggestion that living meant always to be on the high seas, where apart from salvation and doom there would be no solutions, no reservation.'[8] The courage to embark and discover the world is then necessarily connected to the possibility and the metaphors of failure.

The Enlightenment notion of shipwreck did not primarily emphasize the extreme situation of humankind in nature, but challenged science and technology to cope with the intricacies of navigation and protection instead.[9] Roland Barthes, French philosopher and semiologist, portrayed the ship as a symbol of seclusion and refuge from life's raging storms.[10] In a confined space the ship keeps at the traveller's disposal the utmost number of valued objects; the ship forms a singular universe floating amid the violent tempests of time. Barthes particularly referred to the fictional narratives of Jules Verne, in which ships replicate and preserve the world on a small scale. In his novel *20,000 Leagues under the Sea*, published in 1870, Verne describes in great detail Captain Nemo's *Nautilus*, a submarine which includes a library fitted with dark rosewood shelves, inlaid with brass, that hold 12,000 treasured volumes. Moreover, the *Nautilus* contains a valuable collection of art and music and a magnificent museum whose cases display the varieties, rarities and curiosities of nature.[11] Barthes depicted Verne's ships as vehicles of the encyclopedic project of the nineteenth century, designed to encompass and conserve *en miniature* all elements of a finite but rapidly proliferating world.[12]

Appropriating and preserving the world by compiling, registering and neatly arranging the elements within it is a strategy not limited to the modern era of scientific collecting, archiving and interpreting facts. The procedure recalls the primal ship representing the family of mankind, the inventory of the world: the biblical ark. This vessel from the Old Testament (Genesis 1:6–9), furnished with specimens of living creatures on earth, differs in a significant way from Verne's crammed but comfortable floating interiors: Noah's ark is the paradigmatic heterotopia, a storm-tossed place of survival and salvation in the face of catastrophe. In the second volume of his work titled *Sphären*, the German philosopher Peter Sloterdijk analyses the ark as the perfect example of the 'ontology of enclosed space'.[13] *Ark*, from the Latin *arca*, is the word for case or compartment. To Sloterdijk, the ark denotes an artificial interior space, a 'swimming endosphere', that under certain conditions provides the *only possible* environment for its inhabitants.[14]

The archetypal ship reappears in Romanesque, Gothic or Classical church architecture, where the nave embraces the congregation and symbolizes community. The primal ship conveys a strong sense of authority, order and rigorous discipline. But the ark also suggests the uncertain course, the drifting voyage and the eternal journey. Ultimately, it can be seen as an exemplary site of selection and selective principles for survival or obliteration. Whether a ship of ghosts, a ship of fools or a ship of state; whether wind-ridden or painstakingly furnished and controlled, the elements of the primal ship reappear in all ship narratives. The ark embodies all historical and fictional meanings that have been conveyed by ships: enclosure, collection, selection, hierarchy, passage and the upholding of strict boundaries between interior and exterior spaces. A remarkable exposition of these motifs is Arno Schmidt's *Gelehrtenrepublik* from 1957. The novella ridicules the meanings of archive, autarky and authority that the image of the ark embodies.[15] Schmidt recounts the experiences of a young American journalist by the name of Charles Henry Winer who receives permission to visit the International Republic for Artists and Scientists, short IRAS, in 2008. IRAS is an isolated and to a large extent autarkic island that travels the Pacific Ocean like a ship. Fifty years after a nuclear war has devastated most parts of the earth and all of Europe, the miniature republic that mirrors the world on a small scale was created according to international agreement to serve as a refuge for the world's remaining geniuses and their original works.[16]

Naturally, the sanctuary of humankind and paradisiacal reserve to safeguard and cultivate the greatest intellectuals left on earth turns out to be subjected to the same Cold War rivalries that led to the nuclear catastrophe in the first place. Through the eyes of Winer and his diary notes we witness how the familiar system competition is played out among the republic's artists and scientists. The ship/island is equally divided between an American half – the 'free world' – and a Russian half, separated by a neutral area. It accommodates the museums storing all the original paintings and pictures and the libraries containing virtually every book left in the world. In a satirical twist, Winer has us discover that despite their stunning working conditions the creative geniuses turn out sluggish and unproductive. After their two-year probation period they are through with everything but with a new book. The conservationist project is on the verge of becoming an end in itself: 'We are gradually turning into a depot'.[17]

Small Worlds: Castaways, Stowaways, Refugees, Survivors

Blumenberg speaks of the 'nautical arrangement' – of humankind having to come to terms with permanently drifting in the sea, never being able to call at a port. His observation that structures have to be maintained while swimming; repairs and modifications have to be carried out en route, turned out to be an accurate description for the archival ships of the twentieth century, like the ark, the IRAS, the spaceship and also Spaceship Earth.[18] The nautical arrangement involves the acceptance of living with shipwreck and the possibility of a castaway existence, a chance mediated only by increasing control and stability, technology and navigational skills. In terms of Blumenberg's nautical imagery the nineteenth and twentieth century generated an abundance of stories in which the ship signifies a contained environment set against a hostile nature, relating an interior world to an outer world that may be dangerous or promising likewise. Seafaring has been glorified in countless descriptions of remoteness, isolation, tranquility and immersion in a powerful, awesome and sublime nature, of which the breathtaking and fantastical oceans form a vital part.

Just as often, crews saw the oceans as jagged, bleak, eerie, malicious and desolate, as cruelly turning their powers against fragile humans with contempt and inexorability. In the maritime literature of the nineteenth century, descriptions of a homey idyll onboard amid a rebellious and dangerous environment are quite common. Joseph Conrad's novel stresses male camaraderie and fraternalism onboard, and a spiritual order that seems to exclude social inequalities and claims to private property. Writing against the increasing social tensions of the industrialized England of the late nineteenth century, Conrad's seafaring fiction turns the ship into a real and an imagined space in which social realities experienced on land could be suppressed and ignored. Conrad's works thus contributed to the sentimental idealizations of a life of order and solidarity at sea.[19]

Gender is one among many social categories to constitute stable social boundaries in the history of seafaring. Across the centuries navigation has been an exclusively male domain. Both historiography and fictional literature tell of the adventures of tough men on rough seas and their fight against the untamed 'virgin world'. As the ship formed a contained and endangered space, the community on board formed a fragile union, a special maritime culture that Helen Rozwadowski has called a 'small world'.[20] The strict separation of the sexes in turn facilitated certain gendered structures among the crew. The ship needs to be understood as a heterosocial space that did not support one singular masculine self-image or one particular social style; sailors on sea became 'men' in more than one way. It was the absence of women on board, Creighton suggests, that turned gender into a relevant category.[21]

Ships are paradoxical sites where a rigid and hierarchical order of space and conduct prevailed while at the same time equality was cultivated. In expressing a strict hierarchy on board through gender relations, life at sea was romanticized, and discipline and subordination were legitimized and enforced.[22] The solidarity of men constituted itself in relation to the ship that was imagined as female. The community on board was a polyandrous community of men that shared the ship as a common possession.[23] Lillian Nayder analyses the relation of the captain to his ship in Conrad's novels as the relation to a woman that could – in accordance with Victorian images – be either angel or whore. The ship represented a servile asexual figure, an angelic bride or a good soul on the one hand, and a rebellious, demonic female body on the other; nevertheless, it was always seen as in the hands of the male crew. Nayder also shows that the ship was seen as a manmade technology and thus in another way liberated from the image of destructive natural forces. By being manmade, she argues, the ship attested to patriarchal power instead of undermining it.[24]

Faraway and little-known worlds have been connected with the violation of conventions throughout Western literature.[25] The edges of the known world permitted retreat and aberration. The oceans were held to be as treacherous and immoral as the poor, the orphaned, the criminal and the insane, who made them into their sites of refuge. Sailors were from among the socially lowest classes, and crews were often assembled from varied and unlike people. Despite of the strict orderliness and regularity on board mutiny was a ubiquitous threat. In the mid-nineteenth century Herman Melville exemplified this view on the eccentricity of seafaring in his epic *Moby Dick*. Captain Ahab's obsession for the catch of his life in the hunt for the white whale ends in catastrophe for the entire crew. The story is an allegory of the ambition of leader figures and the ruin they can entail.

In numerous literary examples the ship has been portrayed as a site of fundamental dilemmas of life. For instance, Mark Twain's *Adventures of Huckleberry Finn* gives an account of America's struggle with racism in the late nineteenth century through a literary journey down the Mississippi River on a raft. Twain turns a life's story of effort and maturity into a vessel's voyage. Likewise, in the twentieth century Ernest Hemingway's *The Old Man and the Sea* tells of the endless and futile search for the place of the self in the world by highlighting the figure of the boat setting out for the high sea.[26]

Travelling the fragile boundaries between inner and outer worlds involves precarious displacements and transitions. B. Traven's *The Death Ship* tells a story of a lost identity, in which a sailor without papers and unable to prove his name and origin signs on to a ship on which a motley crew of undocumented workers toil away, unaware that the ship will be sunk by its owners to claim insurance.[27] The ship marks the realm between worlds, a place that belongs nowhere and that becomes almost invisible. This is one example of how ships figure as fragile containers, fraught with hopes to reach a better place, in stories of passage. The passage expresses the provisional and suspended character of travelling, the shifting meanings of personal legality in transit. The voyage of the *St. Louis* is another such story of emigration and refuge. The ship left Germany for Cuba in May 1939, half a year before the outbreak of the Second World War, with more than 900 Jewish refugees on board. The hope of finding a haven in another country was dashed when Cuba rejected the passengers. Neither the United States nor any other country came to help. After an infamous round trip, followed closely by the entire world, the journey ended back in Europe, and for many of the passengers it ultimately ended in death.[28]

Yann Martel tells another quite unusual tale of loss and survival in his 2001 novel *Life of Pi*.[29] Pi is a boy from India in the late 1970s who finds himself the sole human survivor of a shipwreck in the Pacific Ocean. In the novel Pi recounts the 227 days he spends on a lifeboat alone with a Bengal tiger, after the tiger has killed the hyena, the orangutan and the zebra onboard. In many ways, Pi's lifeboat refers to the ark, sometimes explicitly: 'You find yourself a great big lifeboat and you fill it with animals? You think you're Noah or something?' But the story is not so much about certain animals to be saved; rather, it addresses the fundamental conditions of survival: 'You can get used to anything – haven't I already said that? Isn't that what all survivors say?' Pi gives his account of life exposed to its bare state, to the point where 'survival' takes its place: 'Life on a lifeboat isn't much of a life. It is like an end game in chess, a game with few pieces. The elements couldn't be more simple, nor the stakes higher. Physically it is extraordinarily arduous, and morally it is killing. You must make adjustments if you want to survive. Much becomes expendable'.[30]

Martel addresses the fundamental questions of selection that seem to be forced upon any lifeboat situation.[31] Ship narratives often impose on the recipient culture a clear distinction of what can and should be rescued and what becomes expendable. The question of what will have to be left behind may be a question of taxonomic completeness, but it may also become a question of life and death. The solution that Pi chooses relies on cross-species kinship. Pi treats the tiger with respect, as an equal, a companion: 'I had to tame him. It was that moment that I realized this necessity. It was not a question of him or me, but of him *and* me. We were, literally and figuratively, in the same boat. We would live – or we would die – together'.[32] Just as often, however, lifeboat narratives insist on the unmistakable choice, on the exclusive either/or of survival that is determined by the drawing of straws or by sheer power and force. This lifeboat ethics of the necessity of selectivity counters the conservative rationality of taking as much as possible on the journey. Instead, the imperative of selection follows a logic of preserving what is practicable or profitable, valuable or useful.

In his story titled 'The Stowaway', Julian Barnes aptly summarizes the themes of authority, order and classification associated with ships.[33] A woodworm that has been stowing away on the ark describes in detail the conditions of the ark's journey that have not been handed down in the books. First and foremost, his narrative concerns the strict discipline and the levels and measures of security in place: 'It wasn't a nature reserve, that Ark of ours; at times it was more like a prison ship'. The woodworm's eyewitness account also straightens up the sentimental myths of the happy couples of the animal kingdom walking aboard. The perspective taken sheds new light on the criteria of selection. It reveals who was eligible in the first place: 'I was never chosen. In fact, like several other species, I was specifically not chosen'. Like most ships, the ark sorted its passengers into separate classes. The category of purity qualified some species, ruling out cross-breeds: 'Some creatures were simply Not Wanted On Voyage'. The anonymous woodworm reminds us that in the selection process for the Compound of the Chosen 'Noah – or Noah's God – had decreed that there were two classes of beast: the clean and the unclean. Clean animals got into the ark by sevens; the unclean by twos'. Referring to the food supply on the ark, where being 'clean' implied that animals could be eaten, a comical picture of a long and dangerous voyage emerges in which Noah manages to deprive the world of its original richness of wildlife.[34]

Modern narratives of ships' odysseys, of lifeboats and of their economies of selection in struggles over borders, limits and capacities seem mostly to address refugee ships crossing the oceans between continents seeking asylum. Nowadays, these migrating ships do not come from Europe in search of a new home in America's lands of promise; rather, contemporary refugee dramas feature refugees leaving Africa heading for the coasts of 'Fortress Europe', most of them, like their fellow immigrants disembarking on Ellis Island, only to find out that on close inspection they are rejected from entering their Promised Land altogether. These events seem to invert the traditional lifeboat narratives; they centre on the security not of those aboard but of those on land. When the rich nations of the world began to limit severely the numbers of refugees they would accept, the ship became the metaphor of the protection and defence of those asked for help but denying refuge. 'The boat is full' became the often-heard statement to keep refugee ships from landing, arguing that the lifeboat they are trying to board is already crowded and provisions will not suffice for all.

Explorers: Mapping Time and Space

In 1952 the US historian Walter Prescott Webb published a book titled *The Great Frontier*. In this book Webb is concerned with what he calls 'the frontier factor in modern Western civilization since about 1500'.[35] Webb refers to a world frontier, or 'Great Frontier', that he conceives of as being European in origin. Western Europe is described by the term 'Metropolis' to signify the 'cultural center holding within it everything pertaining to Western civilization' and comprising the known world, except for Asia, 'which was vaguely known and had no part in this exposition'.[36]

'Then came the miracle that was to change everything'.[37] With these words Webb points to the brief period starting in 1492 in which three new continents were revealed to the European world. The process of European exploration and globalization that was set off by Columbus sailing out in the service of Spain and 'discovering' the New World became known as the European Miracle.[38] The European Miracle does not simply mark the onset of European colonialism; first and foremost, the expression communicates a specifically European history of globalization, that is, a narrative of the world as having been civilized from its cultural centre. The term 'myth' does not equal fiction; rather, in using the term I address the shared narratives at the heart of the collective self-understanding and identification of a culture. To employ the concept of myth analytically allows to combine semiotic and material elements in the formation of reality, and to demonstrate how spatialization, to follow Donna Haraway, involves 'recursive layers of stories and metaphors'.[39] The European Miracle is the mythical narrative that conveys how Europe preceded other world regions by its highly advanced cultural technologies, knowledge and infrastructures, to explain the colonial expansion as inevitable and natural.

In retrospect, a point of origin and a continuous and directed course of development were inscribed into European history. The resulting narrative was teleological in character and legitimized the history of European hegemony after 1492. The retrospective construction of progression, as Sandra Harding notes, also isolated the historical time of central Europe from the times of other cultures. European history stretches from the present to the past in a 'tunnel of time', straight back to paradise.[40] Beyond this tunnel, history seems irrelevant or even nonexistent. Every event considered important for the history of civilization happened in this tunnel; likewise, only meaningful events happened there. The time tunnel is one motif in the mythology of historical progress of civilization. In the teleological perspective, history acquires the authority of making an essential and logically justifiable promise of happiness. Along the story of the European Miracle, Europe created for itself not only a stable past but also a utopian future. The miraculous origin of beginning, what is called the Renaissance, is mirrored in a utopian ending.

How such progressive history and the corresponding homogeneous spaces of geography, as well as distinctive cultural and gender identities were constructed by locating, exploring and crossing boundaries became a key issue for reflection in poststructuralist and postcolonial studies and feminist theories. At the heart of attempts to deconstruct the history of Eurocentrism and the Eurocentric view of world history are 'standpoint epistemologies' that contrast the situatedness of different knowledges with the alleged universality of knowledge. Standpoint epistemologies like Harding's 'Borderland Epistemologies' hold that cultural images of a centre and its margins, associated with the European and the exotic, the civilized and the wild, the male and the female, are Western constructions closely related to, yet concealing, the manifold places in which knowledges have been generated and expressed.[41] Taking the perspective of the situatedness of subjects and their specific knowledges, the master narrative of the Western 'discovery' of the world needs to be understood as a multitude of cultural encounters that took place under conditions of inequality and were often decided by force of arms.

Boundary work has been essential to firmly construct the master narrative of European expansion across locality, difference and disparity. Voyaging Europe made itself up in travel narratives that relied profoundly on exploration by ship. Ships made history, so goes the saying, particularly ships made the history of the European Miracle. From Columbus to Cook, 'oceanic, rather than terrestrial, dominance characterized early modern European empires', historian Elizabeth Mancke argues.[42] Spain, Portugal, Great Britain, the Netherlands and subsequently France became important colonial sea powers. In this Age of Reconnaissance their explorative journeys opened up South America, Central America and the Caribbean, Africa, China, India and Russia for European trade and colonial exploitation. The rising production of knowledge and technologies under European rule seemed to confirm the idea and the history of Western superiority. The 'West' was constituted as a real space of power and as the symbolic centre of global power.

Next to colonial expansion the voyages of discovery constituted the idea of scientific progress as another concept supporting European modernity.[43] The European Miracle paved the way for the miracle of the Scientific Revolution, another transformation that is said to have laid the foundations of the modern sciences in Western Europe in the sixteenth and seventeenth centuries.[44] In part this revolution owes its history to world exploration by sea. Voyages by ship have been central in gaining and storing knowledge on geography, oceanography, botany and astronomy. *Terra Incognita*, the unknown land beyond the horizon, seemed promising to natural philosophers and likewise to conquerors. Major expeditions that were both scientific and colonial in their explorative goals comprised projects like the Magnetic Crusade, the search for the Northwest Passage and the quest for the legendary landmass of *Terra Australis Incognita*.[45]

As naturalists mapped out unknown territories and contributed to creating new empires they also consolidated the conception of absolute space, which was turned into an accepted thought between the seventeenth and the nineteenth centuries. Absolute space relied on the basic metaphysical models of classical mechanics, mathematical realism and scientific rationalism as elaborated in Newtonian and Cartesian philosophies.[46] Within this space, conceptualized as *a priori*, infinite, isotropic, inert and principally empty, territorial power evolved. Earthly space was conceived as uniform, consistent and essentially vacant; with regard to its powerful European centres, the space of the earth seemed to await its progressive filling by Western cultures, people and technologies. Georg Forster, Alexander von Humboldt and Charles Darwin are just the most famous examples of explorers and natural philosophers who extended geographical knowledge, circulated technologies of navigation and cartographical methods, and helped to set up intricate networks of trade relations that also enabled the creation of encyclopedic collections of species and treasures from all over the world.[47]

The knowledge that would become universal was constructed in processes of universalization, of defining what held true and of distributing the principles and standards from the 'centres of calculation'[48] to the places perceived as peripheries. The history of globalization needs to be narrated as a history of accumulating, transmitting and reorganizing local knowledge according to the growing global networks that connected and homogenized the many sites across the earth. The non-European origins of early modern sciences and technologies were habitually made invisible when incorporating the corresponding terrain into imperial territory. Harding has pointed to the discrepancies of the European project of expansion that has successfully blurred the differences between European science and its others, while at the same time it was instrumental in widening the breach between the two.[49]

In the nineteenth century, when states were formed into nations and natural history was disciplined into an array of earth sciences, it was not the outstanding explorer and navigator but the naturalist and scientist who set out for the exploration by sea, followed by nationally endorsed expeditions staffed with crews of earth scientists. Expeditions were seen as a national scientific endeavour for territorial gain as well as a demonstration of scientific prowess and national distinction. The unprecedented grids of scientific observations and precise measurements taken around the world characterized the emerging networks of knowledge. The new networks of scientific observation relied on the growing intricate networks of communication, with telegraphy at the fore, which were often based on imperial occupation.[50]

Thomas Richards argues that in the Imperial Age the affiliation of annexed territory posed the problem of distance: by their mere extension empires tended to evade the final grasp and control. Imperial nations could no longer effect long distance control by pure military force. Empires were 'partly a fiction', 'united not by force but by information'.[51] Since empires posed immense administrative challenges, 'belonging' to an empire was to some extent a fictive affiliation. This attachment, Richard claims, was based on facts that were regarded as the most certain kind of knowledge. Territory was surveyed and mapped, censuses and statistics produced, information accumulated and scattered fragments of raw knowledge were stored, compared and classified. The 'archive' represents Richard's observation that 'it was much easier to unify an archive composed of facts than to unify an empire made of territory'.[52] In his Foucauldian reading of empire building, the archive stands for the attempt to organize all knowledge into a coherent and comprehensive whole. The 'imperial archive' is his term for the 'fantasy of knowledge collected and united in the service of state and Empire'.[53]

Such archival constructions of territorial power were based on technologies of remote investigation, operating in great distance from the centres. Again, ships figure prominently in these histories of remote sensing. Ships have themselves been described as 'instruments': following a thought by Bruno Latour, Richard Sorrensen envisioned ships as part of a larger instrumental entity operating out of a metropolitan capital and reaching across the globe.[54] Leaving tracks, marks and inscriptions on maps, ships became devices of spatial perception. Ships connected colonies to the nations they originated from and yet were heroically remote, linking, in Richards's words, 'the control of territory with a hermeneutics of information'.[55]

In the process of scientifically cultivating global space and of curbing fantasies about unknown landmasses and obscure ocean depths, the relations and interaction of science and popular traditional accounts about strange and fantastic worlds were newly negotiated. The modern travel narratives did not come only in the form of logs and journals, words and pictures; they were written in the languages of physics, geography or botany, structured as tables, graphs and maps, spelled in the form of data by precision instruments, and measured in the units of numbers, lists and specimens. These new travel narratives restructured the view and the experience of the earth. Humboldtian Science became the term to signify a new empirically grounded and instrumentally based scientific approach to derive the order and regularity of nature from close observations expanded across the entire surface of the earth. Humboldt himself introduced his science of the earth as 'terrestrial physics'.[56]

Verne's story of Captain Nemo and the *Nautilus* mediates between the different images of earthly space available by the nineteenth century. On the one hand, Nemo's sublime undersea journey covers fantastic regions like the lost continent of Atlantis, keeping the mythical narratives alive.[57] On the other hand, Verne's story is one about the scientific enlightenment of the depths of the unknown, highlighting scientific precision, technological accuracy and the notion of spatial control of an underwater empire. It is an example of how the modern scientific recreations of the world did not leave popular phantasms behind but entangled the unknown, the fantastic, the ambiguous and the obscure into their accounts, recounting them in new ways and versions. To plot Nemo's journey Verne crossed the Atlantic as a passenger aboard one of the cable ships laying out the telegraph cables on the ocean floor. Verne educated himself with the first comprehensive book on physical oceanography in 1855, *The Physical Geography of the Sea*, arranged by one of the founders of oceanography, the US naval officer Matthew F. Maury.[58]

Throughout the journey of the *Nautilus* around the world, the involuntary passenger Pierre Aronnax is bestowed with historical, geographical and nautical specifics and data listed with scientific exactitude and numerical precision, creating a new style of scientifically informed nature romanticism: 'The Atlantic! A vast expanse of water whose surface covers 25 million square miles, 9,000 miles long, with an average width of 2,700'.[59] The spaces that the *Nautilus* travelled are turned into scientific objects and measured out completely. Employing the latest gear of registering instruments and control devices like the thermometer, barometer, hygrometer, compass, sextant, telescope, chronometer, speedometer, manometer and sounding equipment, the new scientized spaces are navigated. The technology aboard is driven by the powerful agent of electricity, gained by tapping energy resources from the oceans.[60] The issue of control is central to the story. Control refers to the power over space as well as to the command over the ship. Nemo points out to Aronnax: 'I have complete confidence in the *Nautilus*, since I am her captain, her builder, *and* her engineer!'[61]

The *Nautilus* anticipates the spaceship of a hundred years later. It is a hyper-technologized container built to explore its surrounding environment. 'There you have a ship *par excellence*!'[62] These reflections add a new perspective to the history of globalization understood as the narrative of Western superiority over parts of the world 'vaguely known' that had, according to Webb, 'no part in this exposition'. The history of globalization needs to be told first and foremost as a story of imperial and scientific imagination and exploration that made the 'frontiers', the 'horizons' and 'edges' of the known world, into a finite globe.

The Frontier Myth of Recurring Progress

> Europe was crowded; North America was not. Land in Europe was claimed, owned and utilized; land in North America was available for the taking. In a migration as elemental as a law of physics, Europeans moved from crowded space to open space, where free land restored opportunity and offered a route to independence. Generation by generation, hardy pioneers, bringing civilization to displace savagery, took on a zone of wilderness, struggled until nature was mastered, and then moved on to the next zone. This process repeated itself sequentially from the Atlantic to the Pacific, and the result was a new nation and a new national character: the European transmuted into the American. Thrown on their own resources, pioneers created the social contract from scratch, forming simple democratic communities whose political health vitalized all of America. Indians, symbolic residents of the wilderness, resisted – in a struggle sometimes noble, but always futile. At the completion of that conquest, that chapter of history was closed. The frontier ended, but the hardiness and independence of the pioneer survived in American character.[63]

Such are the plot and the main features of the story of the American Frontier as rendered by the historian of the American West, Patricia Nelson Limerick. The term 'frontier', originally meaning border or borderline, obtained new meaning in the course of the settlement of the American West in the eighteenth and nineteenth century. 'The existence of an area of free land, its continuous recession, and the advance of American settlement westward, explain American development'.[64] Frederick Jackson Turner declared his 'frontier thesis' in the year 1893 defending the argument that the centre of American history was to be found at its edges. Turner refers to the frontier as the borderline between white, Euro-American settlers and savage Indians, between civilization and the wilderness that was often termed the 'free land'. Many historians date the beginning of the frontier process with the English settlement of Jamestown, Virginia, in 1607. In the year 1890 the US Census Office declared that the frontier no longer existed; the closing of the frontier, however, was gradual, covering the years from 1880 to 1910.

The story of the frontier reads as a creation myth for American history, as a tale of origin that explains where a group comes from and why it is chosen for a special destiny. Limerick, as quoted above, stresses the features of population, of crowdedness and ownership versus open space and free land. The oft-featured religious motifs of redemption, however, are curiously absent in her version, while generally they play a major role in the story of the frontier. The movement towards the West was pictured as God's providence, as Manifest Destiny. A small chosen people driven out of their homeland were seeking to start from scratch on a new continent, in a New World. The assignment and mission offered a common identity through the shared and renewing experience of violence in confrontation with the Indian 'savages', and it promised the opportunities, the success and the rewards of an independent, democratic community in a land illustrated as the paradisiacal garden of the world.

The narrative of the frontier sustains the image of a dangerous and nonetheless attractive edge, verge or limit to push forward. Frontier areas can be seen as 'remote places' developing their own 'frontier societies'. Similar to colonies fixed to an empire, they are connected to other parts of the world, yet they are distant places 'on the outer colonial edge' of the Old World.[65] More generally, however, a frontier is neither a place nor a geographical line of places but a historical process. 'It is something that lives, moves geographically, and eventually dies', Webb explained in the 'American Frontier Concept' in 1952:[66]

> The American thinks of the frontier as lying *within*, and not at the edge of a country. It is not a line to stop at, but an area *inviting* entrance. Instead of having one dimension, length, as in Europe, the American frontier has two dimensions, length and breadth. In Europe the frontier is stationary and presumably permanent; in America it *was* transient and temporal.[67]

While Webb is infatuated with the pathos of the American Frontier experience and its function in creating democracy and freedom from Old World restrictions, I would like to address the structure of the frontier narrative and its main element, recurrence. 'His [Turner's] most compelling argument about the frontier', the historian of the American West William Cronon argued, 'was that *it repeated itself*.[68] The frontier, as his colleague Limerick stated, 'repeated itself sequentially', 'generation by generation'. The frontier becomes a timeless place, an imaginary site.[69] It is renewed wherever the narrative of a chosen people, driven out of their home and needing to enter and settle a New World, constitutes its collective identity. To cite Limerick again, 'the West is wherever the American mind puts it'.[70] The US space freighter *Valley Forge* in the movie *Silent Running* makes a good example. *Valley Forge* is named after the battlegrounds near Philadelphia where General George Washington endured the hard winter of 1777 with the army of the newly formed United States of America and waged a heroic struggle against the elements and the sinking morale of his soldiers. Valley Forge still ranks as a foundational element in the national historical consciousness of Americans. It signifies a place where by absolute discipline immeasurable suffering and sacrifice will finally be transformed into hope and renewal.

The frontier has moved to semantic areas that are far removed from the original narrative of New World settlement. Nowadays frontiers are imbued with an excess of meaning. Frontier areas are still conceived of as spaces defined and allocated as property by drawing lines on a map, in a struggle to define boundaries on a landscape, forming space into definite places to be added to the body of already known and secured territory.[71] Those boundaries, however, need not be territorial; they can be political, economic or social. Most references to the 'new frontiers'[72] of the twentieth century address the frontiers of knowledge production. Generally, the frontier encompasses any line drawn to create and secure identity by establishing difference. Frontier myths comprise stories of individual and social challenge and adventure, of struggling against, and of overcoming obstacles, and of the rewards of fulfillment of destiny, 'the desirable culmination of the past in the present, laden with future promise'.[73] The frontier is never the obstacle to 'manifest destiny' but its expression.

In 1984 a US magazine confirmed: 'We are a people born to the frontier, and it has not passed away. Our move into space has opened up the greatest frontier of all, the frontier that has no end'.[74] Being 'born to the frontier' became a unique American quality and an immediate expression and renewal of a present-day manifest destiny. 'Space: the final frontier' was a motto of sorts for the starship *Enterprise* in the science fiction TV series *Star Trek* (1966–9), and in 1976 the US physicist Gerard O'Neill published *The High Frontier* on the future of humankind in space.[75] In the 1960s it was the spaceship – scientifically and technologically propelled and sustained – that promised the opportunities and the rewards of a utopian society flourishing in a recreated Eden.[76] The heavens had been compared to a great ocean in the sky as early as the Old Testament (Genesis 1:6–7), and spaceflight could be placed in a direct relation to the expansion of humankind through seafaring. Again, religious motifs figured plentiful as the spaceship formed a direct analogy with the ship, and also with the ark.

Great Frontier – New Frontier – Endless Frontier

The notion of the sciences advancing to transform chaos into natural law and taxonomical order corresponds to the image of time advancing chronologically from the archaic past to a golden future. Both movements are teleological in character and reiterate the frontier narrative. Webb's Great Frontier forms an example of how this relation was perceived and visualized in the mid-twentieth century. Webb superposes the image of the well-known European 'metropolis', the old centre of the world surrounded by continents yet to be discovered and symbolizing the Great Frontier, and a more abstract idea about the generation of knowledge in which the frontier is symbolized by the boundary separating the well-known facts of science from the realm of that which is only imagined and yet to be explored further (see Figure 2.1).[77] According to Webb, the historical process of knowledge production can be described by a model of progressive movement in which in the vast realm of the not-yet-understood is gradually and steadily transformed into positive knowledge (while he allows for an outer realm in both of his model images that refers to the 'unknown', the realm that may never be known, since it is not even known of).

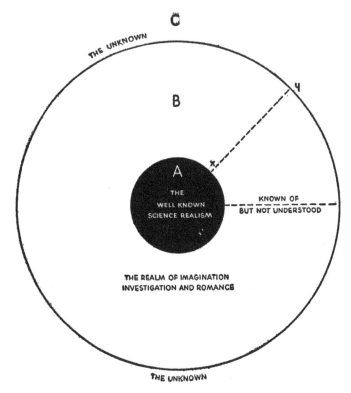

FIG. 1 THE LIMITS OF KNOWLEDGE

Figure 2.1: The Limits of Knowledge. W. P. Webb, *The Great Frontier* **(Boston, MA: Houghton Mifflin Co., 1952), p. 349. Copyright ©️ 1952, 1964 and 1979 by the University of Texas Press.**

Examining his own time, the 1950s, Webb closes: 'The inner circle of accurate geographic knowledge has expanded almost to the outer circle. The area B has been so reduced that little room remains to explore or guess or seek adventure and romance in the unknown'.[78] In the mid-twentieth century the Great Frontier that had been opened with the European Miracle seemed on the verge of closing. The progressive territorial sweep that defined the period from the fifteenth century to the imperial heights of the nineteenth century came to its end. By the early 1960s, even the most isolated and uninhabitable spaces were explored: the polar regions, the ocean floors, the upper atmosphere and outer space. Fairfield Osborn noted in *The Limits of the Earth* in 1953, only a year after Webb published his own book, that the formerly distant 'lost horizon' of the seas and the unknown continents had become familiar territory.[79]

In the rivalries of the Cold War the once infinite and unknown planet earth had come under close military observation. Secrecy and intelligence were major aspects of scientific projects like the International Geophysical Year (IGY) of 1957 and 1958. The IGY promoted placing satellites into earth orbit to assist global observations. Both the US and the USSR tested and launched artificial satellites. Already in 1955, before a satellite was in orbit, the administration of US President Dwight D. Eisenhower sought to implement the policy of free access to space as international terrain that would allow the civilian use of space for satellite observations and grant the right of flight over foreign territory.[80] The Sputnik 'shock' in October 1957 thus came not entirely without warning; rather, it was one effect of the intensified post-war earth and space research that was conveniently framed and justified by programmes like the IGY. Even so, the beeps of Sputnik 1 that could be heard clearly over US territory indicated a new potential of total observation that was not simply embraced in the Cold War constellation but met with suspicion and concealment.[81]

Technologies of satellite remote sensing, of visualization and of communication newly spatialized the world. Next to a vulnerability through exposure these technologies exhibited the relativity of location on the globe, they left the earth with no preferential direction or obvious orientation. Geographically, they allowed for no exceptional position, for no open or free land. As a political image, however, exceptionalism remained strong. In reaction to Sputnik, the Americans put their self-image of freedom and understanding in open contrast to the work behind the Iron Curtain. On 6 December 1957 project Vanguard launched a rocket in full view of the world in an attempt to carry a small satellite into earth orbit. Broadcasting live, the young media of television utterly exposed the US in an unexpected way when the launch ended in a massive explosion.[82]

When in January 1958 a Jupiter missile launched the satellite Explorer I, a tape-recorded message by President Eisenhower that had been sent into orbit was broadcasted all over the world. At Christmastime that year, the world community was to witness the American values of freedom and peace in the form both of scientific-technological imperialism and of liberal-democratic imperialism. The fontiers of territorial expansion were re-established as frontiers of techno-scientific knowledge. Vannevar Bush's report to the president titled *Science – The Endless Frontier* had been released in 1945 as a programmatic policy fostering the idea of American superiority through science and technology. Research and development, according to Bush, provided the new and infinite arena of competition and success that replaced the old frontier of the American West.[83]

It was John F. Kennedy who early in his term of presidency claimed the new frontier for space exploration. In a speech to Congress on 25 May 1961 he called for the US to take a leading role in space achievement and to provide the enormous funds needed to meet the national goal 'of landing a man on the moon and returning him safely to the earth' before the end of the decade.[84] 'What was once the furthest outpost on the old frontier of the West will be the furthest outpost on the new frontier of science and space', Kennedy promised to the people of Houston, Texas, on 12 September 1962 on the occasion of his famous 'moon speech' at Rice Stadium.[85] Half a year after John Glenn had orbited the earth in a Mercury capsule and became a national hero, and just a month before the Cuban Missile Crisis reached its peak, Kennedy's speech was designed to commit the nation to an ambitious space programme. Despite the breathtaking growth of knowledge and of scientific manpower in the age of penicillin, television and nuclear power, the twentieth century also saw open 'the vast stretches of the unknown and unanswered and the unfinished' both of outer space and of contemporary science. Kennedy's new frontier directly referenced the frontier myth of origin of the American nation by recollecting the core elements of this story: 'This country was conquered by those who moved forward – and so will space'.[86]

Kennedy summoned up the history of the first settlers arriving by ship and the subsequent conquering of North America. He referred to the familiar motifs of exploration, challenge, courage, strength, adventure, and also employed the images of seafaring, from the waves of modern innovation to setting sail on the new sea of outer space. The US was to become a 'space-faring nation' exploring a new ocean. Like Sloterdijk's 'swimming endosphere', the spaceship constituted an insular habitat for a small group of humans facing a hostile world that they set out to conquer. Moreover, the spaceship was envisioned as a master craft of technology to function under the harshest conditions,

> made of new metal alloys, some of which have not yet been invented, capable of standing heat and stresses several times more than have ever been experienced, fitted together with a precision better than the finest watch, carrying all the equipment needed for propulsion, guidance, control, communications, food and survival, on an untried mission, to an unknown celestial body.

Kennedy's talk resonates with optimism in relating superb materials, precision technologies and expert collaboration. Moreover, Kennedy envisioned a 'mission' ahead that would surpass human technological excellence in the undertaking of 'carrying all the equipment needed' for 'survival'. The envisioned autarky required placing the journey under God's auspices: 'And therefore, as we set sail, we ask God's blessing on the most hazardous and dangerous and greatest adventure on which man has ever embarked'. The highest possible religious guidance would also justify the high costs and hardships asked of the nation in order to procure the desired rewards. Relying on the familiar motif of national collective sacrifice, Kennedy employs the frontier rhetoric of 'faith and vision', of the sacrifice for a benefit not yet known but confidently envisioned from Columbus to the Pilgrim Fathers.

Myth as Reconciliation of Contradictions

The National Cathedral in Washington, DC, is a monument almost as impressive as Westminster Abbey but much younger. This majestic structure, the world's sixth largest cathedral, was built in Gothic style in the twentieth century and finished only in 1990. The 'National House of Prayer', which actually belongs to the Episcopal Church, identifies itself as embracing all religions. Even the nation's belief in science and technology finds a place in the congregation. Notable among the many beautifully coloured windows is the Space Window (see Figure 2.2). This window stands in deep contrast to more traditional stained glass artworks in churches. A single image stretches across the three-part arched window. Coloured spheres represent planets floating in the deep blue of space. Thin white lines circling the spheres represent the orbits of a spaceship, to signify the smallness of humans in a vast universe.

Dedicated Dedicated in 1974 on the fifth anniversary of the Apollo 11 moon landing, the window contains colours based on NASA photographs taken during the mission. At its centre is a two-and-a-half-inch chip of basalt rock brought back from the moon by Apollo 11 astronaut Michael Collins.[87] One of the ceiling vaults nearby is ornamented with a sculpted lunar surface marked with astronauts' footprints.[88] And it should be noted that among the many grotesque and fanciful gargoyles on the cathedral façade appears the figure of Darth Vader from George Lucas's science-fiction saga *Star Wars*. Because of his evil, villainous character, Darth Vader was placed on the dark northern tower of the cathedral.[89]

The state-sanctioned combination of religion, science and science fiction in the National Cathedral is striking. Since it must be presumed that no element of this cathedral is ironic in nature but that every detail has been chosen with emphasis and gravity, it must be taken literally that humans not only ask God's blessing for the space endeavour but that human space flight has advanced to an expression of religious belief. At the lower edge of the space window is written in capital letters: 'IS NOT GOD IN THE HEIGHT OF HEAVEN?' Space flight, already mastered (science) and yet aspired to (fiction), figures in a larger discourse of supremacy and authority, of an immeasurable cosmos created by God, into which humans inscribe themselves through their civilizing accomplishments, and not last by their scientific and technical achievements.

In the Environmental Age, the spaceship did not serve simply as a metaphor to express contemporary perceptions of earthly problems. Rather, the spaceship informed, structured and aligned the argumentative architecture in the Environmental Age. The discourse of Spaceship Earth was fuelled by common cultural

Figure 2.2: Space Window, Washington National Cathedral. Courtesy of Nathan Bauman, http://nathanbauman.com.

reservoirs shared by progressivists and environmental activists alike. Spaceship Earth opened up a discursive horizon against which certain perceptions and solutions of environmental problems seemed reasonable and became dominant. Spaceship Earth became the core of a mythology. Myths are culturally shared narratives that describe collective human realities meaningfully. Myths are presented, performed, they are sustained through stories, traditions, rituals, images and objects, and consolidated in institutional settings. They reach stability through repetition, through being retold and repractised. Their messages, however, may be not univocal but ambivalent, and this accounts for the fascination myths effect.

Myths are not fictitious stories and they do not stand in opposition to reality. Mythical narratives are not simply 'made up' but 'made', to borrow from Haraway's thoughts on the ambivalence of fiction, construction, and reality.[90] Similar ideas are essential to Judith Butler's work on representation through performative action. Butler stresses repetition and repeated citing in processes of 'materialization' of norms.[91] Mythology formulates comparable ideas of reality formation through repetition, although with a closer focus on narrative coherence.[92] Some twentieth-century works, like the work of Barthes, have pursued the semiological foundations of myth construction, while scholars like Blumenberg have stressed the durability and variability of myth as a stable narrative.[93]

Barthes's semiotic analysis of myth highlights the 'neologism' of a concept and its constitutional function as the core element of a myth; a mythology, he states, needs 'most often ephemeral concepts, in connection with limited contingencies: neologism is then inevitable'.[94] Spaceship Earth can be understood as such a neologism that brings together two different notions from the semantic realms of nature and technoscience: the earth and the spaceship. It creates a new concept and language of *earthcraft* that could become meaningful in the constellation of the Environmental Age. Blumenberg highlights the narrative structure and the function of myth.[95] According to his work, the power of a mythical narrative lies in its combination of traditional elements and time-specific variations of a story. In ever-changing versions, the myth's core elements are combined and intertwined with specific situations. Spaceship Earth used the much older motif of the ship as the 'greatest reserve of the imagination',[96] as Foucault has put it, and expanded it to the modern image of human technological supremacy in space.

Moreover, the power of myth lies in its capability to reconcile controversial arguments: Spaceship Earth represented 'crisis' and 'progress' at once, the two central viewpoints of the time concerning environmental issues. As Barthes put it, 'conceptual neologisms are never arbitrary: they are built according to a highly sensible proportional rule'.[97] The spaceship was a figure that enabled environmentalists and technocrats to argue very different positions on and in the same terms. The mythological perspective clarifies how both optimistic space age cargoists and counterculture survivalists could address the earth as a spaceship. Spaceship Earth offered a resort for either political standpoint.

'Myth hides nothing and flaunts nothing: it distorts; myth is neither a lie nor a confession: it is an inflexion. Placed before the dilemma ... myth finds a third way out'.[98] Barthes discusses the third way out as naturalization: driven either to unveil or to liquidate the concept it presents, myth will 'naturalize' it.[99] Naturalization he identifies as the essential function of myth. The following chapters will attempt to overcome the weak and unproductive opposition of fiction and reality and to assess the 'third way out' that myth offers. I will discuss the naturalized versions of the earth that Spaceship Earth offered, flexing and inflecting the global environmental discourse of its time. Briefly stated, the ways out that Spaceship Earth proposed were the functionally transparent earth system, the earthly lifeboat to select survivors, and the technological earth replica to allow for planetary evacuation and colonization. These versions were neither fictions nor real; rather, they were figurations that became real.

3 CIRCULATION: ECOLOGICAL LIFE SUPPORT SYSTEMS

A Tale of Two Revolutions

In 1999 the journal *Nature* published an introductory article on earth system analysis by the German earth systems scientist Hans-Joachim Schellnhuber of the Potsdam Institute for Climate Impact Research (PIK). According to this article, the movement of discovery and expansion towards the edges of the earth, from the Renaissance to the Enlightenment, has gained a new preferential orientation and direction. The author claims that scientific progress propelled humankind beyond earthly space, into a position entirely dissociated from the limited and earthbound world of the past. Resembling their former gods, humans nowadays stand apart from the earth and look at it from a proper analytical distance. To exemplify the endeavour of '"Earth-system" diagnostics in the twenty-first century' Schellnhuber presents the image of a male scientist opening the earth and dissecting it anatomically. Like a surgeon, clad in a lab coat and surgical mask, the scientist vivisects the earth and analyses its inner structures. The inner structures of the earth are correspondingly represented by way of the scientific iconology of cells and molecules, of particle waves, electromagnetic radiation and material flows. According to this image, earth systems science has turned the planet into a living patient receiving open-heart surgery (see Figure 3.1).[1]

With his 'tale of two revolutions',[2] Schellnhuber refers to the Copernican Revolution of the sixteenth century on the one hand, a time in which the earth was moved out of the centre of the universe and placed in a rather peripheral astrophysical position. The period also marks the beginning of the Scientific Revolution. On the other hand, Schellnhuber points to what he calls the 'second "Copernican" revolution' of the latter twentieth and the twenty-first century, the time which again placed the earth at the centre of the scientific universe. To illustrate this shift of scientific paradigms and perspectives, Schellnhuber offers a second image, a woodcut that he cautiously dates to the Renaissance (see Figure 3.2). This image presents the opposite of today's physician-scientist diagnosing the earth. A sapient male figure, pious and believing but also curious and explorative, succeeds in penetrating the sphere of culture, knowledge and territory that forms his orderly and cultivated world. He breaches the narrow boundaries of his time, the low skies of the pre-Copernican era's Ptolemaic understanding of the heavens, only to discover a whole new and infinitely open universe.

Figure 3.1: Earth System Diagnostics. H. J. Schellnhuber, '"Earth System" Analysis and the Second Copernican Revolution', *Nature*, 402 (1999), pp. C19–C23, on p. C19.

According to Schellnhuber, this image lets the viewer re-experience the 'shock of the Enlightenment as expressed in a (probably apocryphal) fifteenth-century woodcut'.[3] It is quite certain that the picture did not originate in the Renaissance. The woodcut is classified as a wood engraving, fabricated anonymously in the late nineteenth century by the French astronomer Camille Flammarion, as Bruno Weber pointed out in detail in the mid-1970s.[4] As a sixteenth-century woodcut, the picture had long served historiography as presenting the worldviews of the late Middle Ages. Only in the late twentieth century was its origin dated forward to the 1880s. Since then the history of the image has become more complicated. Does it show the rather limited medieval worldviews or does it show the new perspectives gained during the expansive Renaissance era? Or rather, does it show the views of the Enlightenment period and the subsequent era of scientific belief in the nineteenth century?

Figure 3.2: The Shock of the Enlightenment. C. Flammarion,
L'atmosphère: Météorologie populaire **(Paris, 1888), p. 163.**

Most likely the image presents the ideas and imaginations that the twentieth century has developed about these time periods. The retrospective reference to the Scientific Revolution of the early modern period today figures as a parody of the nineteenth century's craving for identity and its obsession with history that Michel Foucault noted.[5] Reiterating the familiar motifs of early modern expansion, the engraving placed the Renaissance and the nineteenth century under a single thematic roof. Pointing to the permanence and continuity of exploration in history writing, the image demonstrates the power of historiography in creating and shaping its object, called 'history', as emerging from the overlapping strata of time and the layers that stories have deposited.[6]

Schellnhuber's images also illustrate the theme of boundaries that has resonated in the myths and metaphors of enclosure and expansion, of limits and horizons discussed in the previous chapter. The work of discovery has mostly been portrayed as a male project at the margins of knowledge and at the limits of certainty, penetrating virgin land. While at the time of the Copernican Revolution these boundaries were considered remote, the nineteenth and twentieth centuries witnessed the rapid closing on formerly distant limits and the transformation of the earth from a vast open space to bounded territories. In this chapter I will look at the 1960s and 1970s, a period that has been studied primarily for the conservationist concerns it raised and the politics of environmentalism it stimulated. I contend that this period also provoked a fundamental shift in the conception of life and confined living space. But in contrast to the pre-Copernican notion of a flat and limited earthly surface a new perspective emerged: the earth of the 1960s could be viewed from the outside, opened up and exposed by scientific means.[7]

Much older images of a female nature being dissected by male science resound in the pictures of the planet exposed in a Faustian fashion to know what the world contains in its innermost heart and finer veins.[8] Holistic concepts of the earth as one single organism to be studied like a living being were combined with the principles of systems sciences and cybernetics, which assessed the earth as a self-contained and self-maintained ecosystem. In the post-war era, ecologists set out to discover, analyse and synthesize the biosphere as a closed space of ecosystems functions. This chapter will explore how the new 'architecture'[9] for the perception of life on earth was created, an architecture that integrated the technological conditions as well as the larger political and cultural ideas, leading to an understanding of life as situated within a closed sphere. The conjunction of the sciences and technologies of the earth with the politics and the cultural images of the earth enabled and limited what was perceptible at the specific historical moment that has been framed as the Environmental Age.

I will discuss this transition by studying the history of the biosphere. This concept of the sphere of all life on earth was proposed in the late nineteenth century to demarcate the realm on the planet in which life could exist. In the first half of the twentieth century, the biosphere was integrated into an ecosystem concept of nature to describe the system of the earth's vital functions. The decade I am interested in, the late 1960s and early 1970s, rediscovered the biosphere in the wake of rising environmental awareness. Environmentalists began to use the term biosphere as synonymous with the threatened (human) environment and an increasingly limited (human) living space on earth. At the same time, ecosystems science reinvented the concept and turned the biosphere into an integrated planetary life support system.[10]

The biosphere of the 1970s can be studied as an example of Peter Sloterdijk's 'ontology of enclosed space' introduced in the previous chapter.[11] Enclosing life into a biospheric space created an interior space or 'endosphere' that offered the only possible habitat for earthly life within the known universe. The biosphere denoted a (human) habitat that faced not only a hostile outside world but was also threatened from within by the alarming economic, social and political developments of environmental pollution, resource exploitation and population growth. Elaborate political programmes were developed to protect and preserve natural resources, and visions materialized to regulate natural and social cycles. The biosphere became a large technoscientific system that was modelled with the aim of optimizing its performance.

Schellnhuber's view on the earth systems sciences at the turn of the twenty-first century can be understood as the result of a development that turned the earth into the largest known ecosystem in the universe.[12] Schellnhuber is not speaking metaphorically when refering to the earth's workings as the Planetary Machinery. The detailed plan of its principles that he provides recalls an engineer's blueprint of a complex piece of technology, an oversized contraption.[13] The earth and world models of the Environmental Age were based on engineering concepts of single elements integrated into complex functional relations and framed by a similar iconology. These earth models proposed that it was in principle, and soon also practically, possible to understand and to control the earth in its totality. Spaceship Earth therefore needs to be discussed not only on the background of the Space Age of the 1960s but also, and even more so, as a metaphor of cybernetic regulation and control of nature and life in the emerging ecosystems sciences.

Endospheric Life: The Biosphere

In September 1970 the ecologist G. Evelyn Hutchinson introduced a special issue of *Scientific American* with an article titled 'The Biosphere' that would later be praised as epoch-making.[14] Hutchinson defined the biosphere as the thin film of the earth's surface and its surrounding atmosphere and oceanic depths, in which life forms existed 'naturally'. He characterized the biosphere as the region in which, first, liquid water can exist in substantial quantities; second, which receives an ample supply of energy from an external source, ultimately the sun; and third, which shows interfaces between the liquid, the solid and the gaseous state of matter.[15]

Hutchinson was not the first to restrict life to a 'terrestrial envelope' roughly 20 kilometres in extent.[16] In a brief passage at the end of his book on *Die Entstehung der Alpen* (*The Genesis of the Alps*) published in 1875, the Austrian geologist Eduard Sueß, a professor at the University of Vienna, had reflected on the 'zone' on 'this big celestial body formed by spheres' to which organic life was constricted; 'on the surface of continents', Sueß asserted, 'it is possible to single out a self-contained biosphere'.[17] Sueß introduced to geology the concept of 'spheres', referring to the concentric 'envelopes of the earth' from the lithosphere and the hydrosphere to the atmosphere.[18] Life, however, he identified as being present in a particular sphere. In the final volume of his three-volume *Das Antlitz der Erde* (*The Face of the Earth*), published in 1909, Sueß asserted that 'all of life' arranged itself into an 'entirety' that emphasized not 'unity' or 'common descent' but 'solidarity'.[19] Natural philosophers like Jean-Baptiste Lamarck and Charles Darwin, who held that heredity and evolution followed natural laws, had opened up this notion that brought with it, Sueß argued, 'the concept of a biosphere through which life was assigned its place ... and which at the same time encompassed life only on this planet with all its demands as to temperature, chemical composition etc. and excluding all hypothetical ideas of possible life on other celestial bodies'.[20]

Sueß considered life 'ruled' by the particularity and destiny of the planet.[21] He endeavoured to show 'in which way life nestled up against', or adapted to, 'the face of the earth'.[22] Hutchinson in 1970 instead addressed the Biosphere – with a capital B – in the terms of global geochemistry laid out by the Russian mineralogist and biogeologist Vladimir Ivanovich Vernadsky of the University of Moscow. Vernadsky's main work *Biosfera* was published in 1926 in Leningrad; only in 1998 did it become available in an unabridged English translation titled *The Biosphere*.[23] Vernadsky directly referred to his predecessor Sueß, whom he had also visited in 1910, when rephrasing Sueß's idea of the biosphere as 'a specific, life-saturated envelope of the Earth's crust'.[24] Vernadsky identified three characteristics of his Biosphere: first, that life had its place on a *spherical* planet, derived from the observation that the earth constituted a closed sphere; second, that life was not merely an effect of geological conditions but also a geological force in itself, producing geological effects; and third, that the Biosphere was subject to temporal change, since it continuously incorporated new parts of the earth.[25]

Vernadsky's biosphere was ultimately better acknowledged and received than Sueß's biosphere. Vernadsky's project of a 'physics of living matter'[26] within the parameters of a spherical planet provided a much more powerful heuristic base to study energy and matter cycles within the environment than Sueß's geographical concept. Framing the biosphere not only as a closed container of life but also describing its forces, mechanisms and functions promoted a scientific concept and a modelling of life in ecosystems terms; moreover, this view led to a translation and reinvention of Sueß's 'selbständige Biosphäre' as 'self-contained' as well as 'self-maintained', stressing the self-organizing and self-sustaining power of the biospheric metabolism.[27] Biosphere scientists today speak of the Vernadskyan 'revolution', comparing Vernadsky's formulation of a unified biosphere theory to Darwin's theory of evolution: 'What Charles Darwin did for all life though time, Vernadsky did for all life through space'.[28]

Vernadsky has been praised as a 'cosmic prophet of globalization'.[29] The description is accurate insofar as Vernadsky anticipated the transition of the environment towards a closed space to be assessed in the terms of ecosystems science that would in the late twentieth century entail a fundamental reconceptualization of earthly living conditions. 'From the very beginning', Jeremy Rifkin commented in 1991, 'the worldwide enclosure movement narrowed its frame of reference to the geosphere, the solid surface of the planet. Even after the other earthly realms were invaded, enclosed, and commercialized, the thinking of the age remained fixed on the horizontal plane'. Rifkin characterized the merging movements of expansion and limitation of life to one single, global living space or habitat in the late twentieth century using the image of a dawning 'Biospheric Age'.[30]

The 1970s saw the conceptual transition from a phenomenological understanding of the envelope of life towards a physical and geochemical approach to living matter and its inert environment as a self-regulating, self-sustaining and evolving system.[31] Hutchinson in 1970 based the 'operation' and the 'day-to-day running' of the biosphere on a mechanical and functional idea and terminology of managing an 'overall reversible cycle' (see Figure 3.3).[32] His notion of the biosphere as a metabolism, a not incredibly efficient circulatory system, whose boundary conditions limited the 'amount of life' on earth, made Hutchinson conclude in respect to its expected lifespan: 'It would seem not unlikely that we are approaching a crisis'.[33]

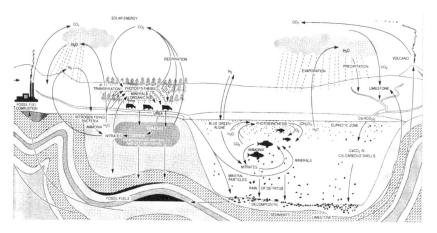

Figure 3.3: Major Cycles of the Biosphere. G. E. Hutchinson, 'The Biosphere', *Scientific American*, **223 (1970), pp. 45–53, on pp. 50–51. Courtesy of the estate of the artist Bunji Tagawa.**

Whole Earth

Both Sueß and Vernadsky concluded that the state of the biosphere was temporally contingent: 'The claims' of life on earth, according to Sueß, 'presuppose that the biosphere is a limited phenomenon, not only with regard to space, but also with regard to time'.[34] This notion of a narrow time-space for earthly life to inhabit supplied a framework and basis for the rational scientific understanding of life processes on earth. Yet this notion also fuelled the perception of environmental crisis that was prevalent in the late 1960s and early 1970s. The special issue on 'The Biosphere' appeared in the thick of an emerging environmental movement in the Western world that replaced traditional nature conservation movements and reprehended the period of post-war economic growth that had relied extensively on science- and industry-based innovations and that had propagated an unproblematic view of science and technology.[35] The ecosciences were a part of the criticism and the alternative views spread. The *Scientific American* special issue took up the dominant concepts and concerns of the time: 'population growth', closely related to 'earth's finite resources', and 'the finite dimensions of the planet', which were identified as the cause for the rising pressure of economic and ecologic 'limits'.[36]

'The Environment has just been rediscovered by the people who live in it', wrote environmentalist Barry Commoner as he mocked his fellow Americans in his 1971 book *The Closing Circle*. 'In the United States the event was celebrated in April 1970, during Earth Week'.[37] 22 April 1970 was proclaimed the first 'Earth Day', a day of nationwide public demonstrations. To express environmental concerns and limits, referring to the earth's totality became the most impressive means of demonstration. Pointing to the planet earth as a single entity stressed its fragility and transience, and its quality as a unique treasure, thereby calling for political unity. Numerous activist groups carried the planet, the world or the earth in their names, like the World Wildlife Fund (WWF), founded in 1961, or Friends of the Earth, founded in 1970.[38]

In 1968 Stewart Brand published the first issue of the *Whole Earth Catalog*. The guidebook promised 'access to tools' for the counterculture in the style of a mail order company.[39] The 1969 autumn issue of the catalogue featured Blue Planet on its cover, a photographic image of the earth that had been taken only months earlier by extraterrestrial photography. By the end of 1967 the first satellite exposures of the earth had been available. It was Apollo 8, the first manned mission to the moon, that broke free of earth's gravity for the first time in December 1968. Orbiting the moon, the crew brought back photographs that, when turned 90 degrees to the right, showed the earth 'rising' over the moon's surface. The Apollo 11 mission carried through the long-planned manned moon landing in July 1969. A worldwide audience watched its transmitted views of home planet earth from outer space live on television.[40]

Historians almost unanimously regard space activities as a salient battlefront of the Cold War. When competition turned into a veritable (arms) 'race', the Space Race formed a prominent part. Notwithstanding its persistent tenor of a world united and truly one, space flight stressed not globalism but nationalism. The images of 'Full Earth' were taken and deliberately signed with national authority. The underlying issue of authorship addressed the question of who aspired and deserved control of the earth on display, and also the question of which nation would be able to launch a space programme powerful enough to provide not only images but also prove its scientific and political command of the earth. Particularly the project of 'manned' space flight became an overwhelmingly national endeavour, expressed, for instance, in the crucial moment of planting the American flag on the moon, possibly the most-featured act of every Apollo moon landing.

US space flight took place at the height of US involvement in the Vietnam War and was also a national project with regard to ongoing international conflict. The astronauts' view of the earth, so small and beautiful, however, could not hide the international and also blatantly national inequalities and disparities of the time. In 1969, The Last Poets in a song titled 'Whitey on the Moon' cynically exposed class and race inequalities in the United States by pointing to the outrageously high costs of the manned space programme and to the injustices of national resource allocations: 'No hot water, no toilets, no lights (but Whitey's on the moon)'.[41] This and similar critical voices spoke to the striking selectivity according to national affiliation, race, class and gender in the question of who would represent humanity in outer space, and of who would have to stay back on a deteriorating planet.

Primarily, the images of the 'Blue Marble' were received as a message of transcendence, of a planet without borders. 'Whole Earth', as Joachim Krausse remarked, obliterated the political boundaries, which the cartographic model of the globe, with its competing lines and letters, had stressed.[42] The images focused on the specific combination of water and air, soil and plantlife that formed the biosphere. This view of the earth from far beyond zoomed in on its distinct features, illustrating Schellnhuber's argument about the Second Copernican Revolution: the totalling gaze from infinity was also partial since it focused on the conditions of life that were held to be unique within the universe.

When first orbiting the moon in December 1968, the crew of Apollo 8 contrasted the barren and desolate lunar landscape with their first view of the beautiful blue earth and its white swirling clouds. 'To see the earth as it truly is, small and blue and beautiful in that eternal silence where it floats, is to see ourselves as riders on the earth together, brothers on that bright loveliness in the eternal cold – brothers who know now they are truly brothers', the American poet Archibald MacLeish wrote in the *New York Times* on 25 December 1968. Reflecting on mankind's changing conception of itself MacLeish remarked that mankind for the first time in all of time had now *seen* the earth, seen 'the life on that little, lonely, floating planet: that tiny raft in the enormous, empty night. "Is it inhabited?"'[43] As Apollo 8 entered lunar orbit on Christmas Eve 1968, the three astronauts, Frank Borman, James Lovell and William Anders, took turns to read out in full length the passages on creation from the Bible's Book of Genesis:

> In the beginning God created the heaven and the earth. And the earth was without form, and void; and darkness was upon the face of the deep. And the Spirit of God moved upon the face of the waters. And God said, Let there be light: and there was light. And God saw the light, that it was good: and God divided the light from the darkness ...
>
> And God said, Let the waters under the heavens be gathered together unto one place, and let the dry land appear: and it was so. And God called the dry land Earth; and the gathering together of the waters called He Seas: and God saw that it was good.[44]

The astronauts' biblical rendering of the creation of the conditions for life on earth, broadcast to the world, shows that next to gender, race, class and nation, human space flight was also marked by religion, which had a firm place and a distinct say in the venture. The message that earth was the creation of a Christian God, placed at the centre of their space voyage, demonstrates that religious belief was never considered to oppose the high-end scientific venture of space flight but was understood as its incentive and its underlying mission. The United States presented itself as the righteous and legitimate performer of the mission of exploring the frontiers of the unknown. Commemorating the country's humble origins formed part of the story of a larger purpose and served to mediate the controversial undertaking of human space flight. The narrative presented the displacement from the original home as an assignment designed by a superior power for superior reasons, which would only disclose themselves with outstanding sacrifice. It is a paradox of such frontier missions that they tend to justify the sacrifices not only with the accomplishment of some faraway goal but also with the reverence to what is left behind: 'The vast loneliness is awe-inspiring and it makes you realize just what you have back there on Earth', said Lovell.

Spaceship Earth

Focusing both on the fragile living conditions on the earth and on the exploration of new worlds, the Space Race connected well with the 'race for survival' of its time. In 1966 the British economist and political scientist Barbara Ward took up the image of the earth as a tiny illuminated Blue Dot floating in the sheer infinite black sea of space to describe the delicate political situation of the world:

> In the last few decades, mankind has been overcome by the most fateful change in its entire history. Modern science and technology have created so close a network of communication, transport, economic interdependence – and potential nuclear destruction – that planet earth, on its journey through infinity, has acquired the intimacy, the fellowship, and the vulnerability of a spaceship.[45]

'In such a close community, there must be rules for survival'.[46] Ward's book, which carries the simple title *Spaceship Earth*, was set up as a challenge for world policy to restore the balance of power between the continents, as well as a balance of wealth between North and South and a balance of understanding and tolerance between East and West. 'In fact', Ward continues, 'I can think of only one way of expressing the degree to which interdependence and community have become the destiny of modern man':[47]

> The most rational way of considering the whole human race today is to see it as the ship's crew of a single spaceship on which all of us, with a remarkable combination of security and vulnerability, are making our pilgrimage through infinity. Our planet is not much more than the capsule within which we have to live as human beings if we are to survive the vast space voyage upon which we have been engaged for hundreds of millennia – but without yet noticing our condition.[48]

Ward published her book at the apogee of the space narrative. While the 'vast space voyage' of the earth had been ongoing for hundreds of millennia it had not been noticed before and much less been voiced in the terms of space flight. Fashioning the human condition in terms of a crew hoping to survive in a tiny space capsule could become meaningful only in the Space Age. The remarkable combination of scientific-technological and political security and vulnerability Ward addresses could not be represented better than in the motif of Spaceship Earth.

'Spaceship Earth' quickly gained currency in different countries and languages, expressing the growing interconnectedness, interdependence and affectedness of the world. The figure became a symbol of the earth's wholeness, of the uniqueness of life and the uncertainty of its conditions, and of the limitedness of its allotted space and time. Spaceship Earth was the icon of numerous programmes on how to gain benefits from a small and scarce world, of ideas and programmes for alternative, more modest and more sustainable standards of living, particularly in the West. The economist Ernst F. Schumacher acquired reputation for his plea for nonviolent small-scale or intermediate technology, or 'technology with a human face', neither centrally nor hierarchically organized like the fossil and the nuclear energy industries and other large-scale scientific-industrial systems. In his book *Small Is Beautiful* from 1973 Schumacher employed the spaceship to reflect on 'a lifestyle designed for permanence':

> A businessman would not consider a firm to have solved its problems of production and to have achieved viability if he saw that it was rapidly consuming its capital. How, then, could we overlook this vital fact when it comes to that very big firm, the economy of Spaceship Earth and, in particular, the economies of its rich passengers?[49]

Spaceship Earth did not appear just as a neo-romantic image of the first global-minded environmentalists. Spaceship Earth also figured in proposals for a technoscientifically managed economy of nature. Rational analytic approaches to the inventorying, stabilizing and optimizing of earthly resources were to provide the 'blueprint for survival'. How these parables of doom and survival partly conflicted and partly corresponded will be discussed in the following sections.

Parables of Doom

> This space voyage is totally precarious. We depend upon a little envelope of soil and a rather larger envelope of atmosphere for life itself. And both can be contaminated and destroyed. Think what could happen if somebody were to get mad or drunk in a submarine and run for the controls. If some member of the human race gets dead drunk on board our spaceship, we are all in trouble. This is how we have to think of ourselves. We are a ship's company on a small ship. Rational behavior is the condition of survival.[50]

Ward uses Spaceship Earth as a metaphor to address the shared fate of humankind in a small world, in which disease, warfare, the nuclear threat and resource scarcity affected virtually everyone, everywhere. Her concerns borrow both from the technologies of space flight as well as from the ecological discourse of the biosphere as the limiting envelope for life on earth. Within a decade of the publication of her book, the 'age of ecological innocence' would come to an end. When oil became costly in 1973 and 1974, alarmist and partly apocalyptic views of the earth took hold.[51]

Risk and security feature prominently in Ward's argument. She elaborates on how the fate of the planet depended on keeping uncertainty at bay. Resource security and environmental security supplemented the global issues of military and energy security at a time when environments were thought to be more fully under human control than ever before. The notions of human power over nature and of its modification have been described as a veritable 'push-button' mentality towards the earth. Cold War projects of weather control or of melting polar icecaps are examples of contemporary ideas of planetary management. Most closely, however, the power of technological intervention was associated with the push-button destruction of the world. The 'technological fix' was closely related to the technological 'unfix', with the risk of the earth's total annihilation.[52]

Historically, the technological potential for intervention seemed unprecedented; the multiple obliteration of the earth by atomic power or biological warfare seemed possible and probable. In this regard, the rational behaviour Ward called for was a double-edged sword. Consistent with the inherent problems of technological control, a language of crisis and alarm prevailed in environmental and antiwar discourses. Both 'genocide' and 'ecocide', the absolute, final, earth-spanning deaths of humanity and nature, were instigated by the latest means of science and technology.[53] As Ward points out, the interventionist and highly rationalist options on a planetary scale allowed both the command of the earth and its obliteration in total:

> For the first time in human history, a nation can lob a little device over a neighbor's backyard and blow him up and everything else with it. If this fact does not create a 'community', I do not know what can.[54]

Around 1970, a vast amount of fictional literature and films imagined genocide, the total annihilation of humankind by nuclear warfare, biological warfare and earth-spanning diseases. One little-known piece of post-apocalyptic ship and island science fiction needs to suffice as an example of how a survivor community was envisioned in relation to Western civilization and modern technoscience at the nightfall of the known earth and the dawn of another. *The Noah*, directed by Daniel Bourla and released in 1974, blends with the Robinson motif of the lone castaway who in the wake of the shipwreck of humankind is washed to the shores of a devastated earth.[55] The black-and-white film is set in the near future of the mid-1970s. It opens by quoting liberally from Genesis 6, the passage of the Bible that describes how man grieved God so that God repented against his creation and decided to destroy humankind and eradicate every living substance from the face of the earth. 'And Noah only remained alive ... '

Noah turns out to be a veteran soldier of the US Army. He disembarks from a lifeboat in the form of an inflatable army raft on a shore in South East Asia, wearing full combat gear. In what follows, Noah colonizes the newfound land by inhabiting a shack on the overgrown grounds of a former army base. The trivia and the posters of communist leaders left in the barracks and the battalion of Mao Zedong busts indicate that he arrived on a deserted base of the People's Liberation Army. These may be the long-forgotten relics of China's involvement in the Korean War of the early 1950s, but more likely we are witnessing a time past the Chinese Cultural Revolution. Most likely, these are the remnants of the Vietnam War in the 1960s and early 1970s. At the peak of the Sino-Soviet split of the early 1970s, China established closer relations with the US, a situation that would explain why the soldier raises and salutes the flag left on the premises.

The soldier is a disciplined warrior of 'this man's army'. He is neither a cynic nor a hero, but a well trained, devoted regular. We hear his alarm going off at six in the morning and watch him dutifully toiling away to the cheerful marching tune of John Philip Sousa's 'Washington Post', performing his soldierly rituals, ardently inspecting his supplies, neatly stowing away his clothes in his locker, cleaning his boots, sweeping his barracks, doing his morning exercises and building a field latrine. Scrupulously he peruses his first aid and survival guidebooks. But silence and isolation soon affect him and his motivation wanes. Patiently he awaits orders that never come – a twisted emulation of Robinson stranded on his deserted island and waiting for the ship that never comes.[56] To cope with his loneliness the castaway invents a companion, an army private with the name of Friday. From the conversation between the soldier and his private we learn that the earth is indeed devoid of humans. While our soldier was in the shelter 'taking inventory', his comrades were surprised by a nuclear strike and its aftermath. This soldier is the last and the first man on earth: 'just call me Noah'. Noah is in command now. He plans to build a boat and leave for the United States of America 'as soon as ... [he] finished taking inventory'.

Noah's obsessive inventory management finds its example in Robinson Crusoe who, struggling to create a new life for himself in the wilderness, takes detailed inventory of the island in the form of lists of what he possessed, of what he did not possess and of possible substitutes. Likewise, Noah imagines himself as the founder of the new earth, and in the course of his projects relives the canonical Western history of civilization until the late twentieth century, mediated by voice and noises only. In his story of remaking the world from scratch, Noah takes up every patriarchal role conceivable. He is God the father, creating the first people on earth by making a female companion for Friday. When the couple inevitably falls from grace Noah refashions himself as a figure of authority and moral by deciding to supervise and teach a school class of children that he can more easily control. Up to graduation, he reviews the entire Western canon of knowledge: mathematics, history, literature and the spirited versions of US liberalism, democracy and capitalism. From his fantasy eventually springs an entire city with all its major institutions: the school, the church, the entertainment industry (the sports arena, the cabaret, the fair, the races), and the military. Out of growing diversity, poverty and discord the known world of the twentieth century arises, with its developing regime clash between communists and capitalists and their armed conflicts.

Noah struggles hard to stay in charge of his creation. We watch him reading Dale Carnegie's guide on *How to Win Friends and Influence People* (first published in 1937). Next we see him figuring as the annoyed Moses coming down from the mountain to lecture his people with tablets containing a set of basic simple rules to restore law and order. 'Now they think I'm a God', he muses. But at last he states that if the choice were put to him he would 'take the army, any time' – 'no backtalk, no arguments, no responsibilities, no headaches, just obey your orders, do what you are told'. In his frustration Noah does what he does best: he devotedly clings to the Western idea of justice, peace and happiness. He invents himself as the commander-in-chief of a large army corps to lead into a long war. By means of contemporary media coverage – a cacophony of music, voices and speeches from an original soundtrack of radio and TV – we live with him through the conflicts and developments in the second half of the twentieth century, from the *Blitzkrieg* to Stalingrad, from Pearl Harbour to the atomic bomb, from Seoul to McCarthy to Joan Baez to Saigon, and finally to nuclear holocaust, to 'cleanse the world of ancient evils, ancient ills'.

Quoting from a declaration by Franklin D. Roosevelt during World War II, the story takes on a sarcastic note, demonstrating how global nuclear war became possible and acceptable by discursively framing the nuclear strike as the apocalypse, as the Judgement Day inevitably springing from the need to purge the world of its dark powers and evils. The film illuminates the close connection of the dark and the glorious in collective Western imagination that justifies the devastation brought about by technoscientific hubris. It exposes the religious zeal, the spiritual fervour developed by a community that considers itself in a state of emergency and in need and capable to decide who will have to die and who may survive. And the film criticizes the repeated reference to a divine mission that authorizes – practically desires – the end of politics. *The Noah* is the parable of the fallacy of man: in due course, we watch Noah the patriarch and the commander-in-chief fall back to his original state. We see a sick and defeated human, about to die from the exposure to radiation in heavy poisonous rain. The global dynamics of the earth's biosphere are about to disrupt his insular condition.

Survival: Taking Inventory

The new global earth of the twentieth century made it more difficult to conceive of any site in the world as a remote island. Yet globality enforced the notion of the earth itself being a small and isolated island on which open space was limited. 'Gradually ... man has been accustoming himself to the notion of the spherical earth and a closed sphere of human activity'. When in 1966 the US economist Kenneth E. Boulding suggested to think and act in terms of the 'closed earth' on the occasion of a conference on 'environmental quality' in Washington, DC, he referred to a historical 'transition from the illimitable plane to the closed sphere'.[57] Reaffirming the arguments of Walter Prescott Webb's *Great Frontier* discussed in the previous chapter, Boulding dated the notion of the 'global nature of the planet' to a time after World War II when rocketry and space flight were in their early stages.[58] His text reiterates the classical topos of the American Frontier and the World Frontier: Earlier civilizations, Boulding stated, had had the experience that there 'was almost always somewhere beyond the known limits of human habitation', so that 'there was always some place else to go when things got too difficult'.[59]

According to Boulding, the 'closed earth of the future' required a new economy. The 'cowboy economy', the open, throughput-oriented economy of the illimitable plains and prairies, would have to be superseded by a 'spaceman economy', a cyclical economical system of the closed sphere capable of continuous material reproduction, and externally sustained by energy inputs only:

> I am tempted to call the open economy the 'cowboy economy', the cowboy being symbolic of the illimitable plains and also associated with reckless, exploitative, romantic, and violent behavior, which is characteristic of open societies. The closed economy of the future might similarly be called the 'spaceman' economy, in which the earth has become a single spaceship, without unlimited reservoirs of anything, either for extraction or for pollution, and in which, therefore, man must find his place in a cyclical ecological system which is capable of continuous reproduction of material form even though it cannot escape having inputs of energy.[60]

'Spaceship Earth' was Boulding's programmatic figure to project a new economy for future earth. His 'spaceman economy' posed the challenge of rationally economizing and managing the entirely closed and circulatory material flows and thus assuring 'stock maintenance'.[61]

'Did you take inventory?'[62] The soldier who invented himself as Noah feared that his superiors would eventually confront him and remind him of his essential mission. His limited world was defined by strict order and by a scrupulous care for his equipment and supplies. His obsession with taking stock can be perceived as a contemporary allegory of the earth and its inhabitants. While there is a tragic component to the very naivety, the simplicity of the compliant effort, there is also a good portion of presumptuousness involved in the conviction of being able to master the challenge. Facing the overwhelming endeavour of inventorying the earth for future improvement is also a reputable, even heroic move. In dealing with these conflicting aspects of taking and maintaining stock for survival, I am interested in the meanings and the consequences of accounting of the earth and for the earth, from the Noah-like inventions to the interventions, from the inventory to the planetary machinery of what had been termed Spaceship Earth. Boulding's oppositional reference to the 'cowboy' ideology introduced Spaceship Earth as an alternative to reckless expansionism. But the reference to space flight also made Spaceship Earth its continuation.

'Time is running out fast for "spaceship earth."'[63] Thus reads the back cover of *Planet in Peril? Man and the Biosphere Today* from 1972. In this UNESCO booklet the US ecologist Raymond Dasmann developed a scenario from the 'impending dangers' of his time, relating the discourse of urgency and sacrifice to the newly perceived dimensions of the biosphere. 'The uncontrolled growth of our species acts as a sort of cancer on the biosphere', Adriano Buzzati-Traverso, UNESCO's Assistant Director-General for Natural Sciences, stated in the foreword. He identified the impact of mankind on the biosphere to be global in scope, growing at an exponential rate, and at a point where the 'entire system which has made life – and particularly human life – possible' was fundamentally affected.[64]

By placing environmental problems into the confined space of the biosphere, the UNESCO booklet accentuated the new dimension of limitedness and finiteness of the earth. In several photographic images Dasmann documented the contemporary conditions of life on earth. Black-and-white photographs were placed directly next to one another and bluntly exposed the shifting spatial relations and the new precarious neighbourhoods on the planet. Juxtaposing the pure and virgin Antarctic ice shelves or the dry African savannahs and the urban industrial outgrowths of the North, Dasmann vividly pointed out that places that had formerly seemed far apart or far remote had moved increasingly close together and were ever more threatened. The terms environment and biosphere were used more or less synonymously. It was the biosphere, the earthly 'living space', that was in danger.

When the first 'biosphere conference' took place at UNESCO headquarters in Paris in September 1968, 'qualitative and quantitative living space requirements' formed its core.[65] The first large international 'conference of experts' convened to solve 'the problems of the biosphere in its totality'[66] and to develop, as the proceedings claim, a 'scientific basis for rational use and conservation of the resources of the biosphere'. Scientists representing twenty-five nations formed the International Coordinating Council, which first came together in 1971 to prioritize and coordinate research within four areas that concerned natural, rural and urban-industrial environments and their pollution. The conference proceedings linked the 'living space' of the biosphere and its increasing insufficiency to resource security and framed the results in terms of 'The Spaceship Earth'.[67] This spaceship was to be inventoried by rational scientific method. Ten pre-studies, prepared as a thematic background for the conference, projected programmes for the future relations of 'Man and his ecosystems'; they followed 'the aim of achieving a dynamic balance with the environment, satisfying physical, economic, social and spiritual needs'.[68]

'The basic concern of this conference is the use by man of his living space on Earth'.[69] The Sueß-like view of the biosphere 'as that part of the world in which life can exist' was further specified as the 'human living space'. A model of a resource-shaped natural world available for human use became the object of international research and politics.[70] The studies focused particularly on vegetation, land and land use; on water and aquatic resources; on animal ecology and the management of wild game; on the threat to natural diversity; and on waste, pollution and health. The environmental crisis was made accessible as a problem of 'the biosphere as a whole'.[71] To study biospheric processes, the UNESCO Biosphere Conference launched the United Nations Environmental Programme (UNEP), which took up work in 1972.

The conference also launched a worldwide research programme under the title Man and the Biosphere (MAB).[72] MAB innervated environmental research by designating 'biosphere reserves' in a coordinated worldwide network that was conceived as representing the ecosystem of planet earth. This new form of protected area reflected the internationalization, but also the integration, of Western scientific approaches to nature, and a move towards global environmental governance in the second half of the twentieth century.[73] The selection of sites, the designation of areas as well as their management departed from the tradition of protecting nature's insular scenic beauty. Biosphere reserves answered primarily to scientists' visions of a global ecosystem management. They represented ecological functions in the global ecosystem and were scientifically selected to conserve 'the diversity and integrity of biotic communities of plants and animals within natural ecosystems'.[74] The International Coordinating Council of the MAB programme adopted thirteen major project areas for study, focusing on different ecosystem types, like forests, deserts or mountain areas, on the conservation of resources like genetic information, or on certain development impacts, like urbanization, land use and pollution. In 1976 the first fifty-seven biosphere reserves were established.[75]

Biosphere reserves protected nature according to the contexts and needs of 'present and future human use'.[76] They represented a new epistemological framework to understand the earthly environment and the position of humankind within this environment. Reserves were tourist sites and places of scientific education and research, but they were also 'living laboratories' to explore the wealth of nature for human benefit and future utilization. Reserve management focused explicitly on the reconciliation of diversity conservation with social, cultural and economic development. This new form of environmental stewardship marked a transition from a nature-centred view of conservation to a human-centred view on the environment.

The international conference of the United Nations in Stockholm in 1972 on the 'human environment' was the first in a series of 'earth summits' at which the international community up to the present day assures itself of holding global responsibility.[77] The Stockholm conference was seen as an opportunity to take stock of the state of the environment. It was also an attempt to achieve international consensus on the threats to the environment. A study coauthored by Barbara Ward and the French microbiologist and environmentalist René Dubos prepared for the conference agenda. In line with Boulding's thoughts, the authors referenced the closing world frontier: 'Now that all parts of the globe are occupied, the careful husbandry of the earth is a *sine qua non* for the survival of the human species'.[78] They invoked the risks for humankind relying on the closed biospheric space: 'We are indeed travelers bound to the earth's crust, drawing life from the air and water of its thin fragile envelope using and reusing its very limited supply of natural resources'.[79] The fundamental task they describe as 'to formulate the problems inherent in the limitations of the spaceship earth'.[80]

Ward's and Dubos's study titled *Only One Earth* had the familiar ring of alarmism to it. However, the subtitle *The Care and Maintenance of a Small Planet* invited it to be read like a technical reference manual. The authors demanded a new kind of environmental action, a combination of environmental empathy and efficient resource management: 'Now that mankind is in the process of completing the colonization of the planet, learning to manage it intelligently is an urgent imperative. Man must accept responsibility for the stewardship of the earth'.[81] Ward and Dubos brought forth the notion of a third party for whom and in whose name mankind in the 1970s was supposed to act: 'The world *stewardship* implies, of course, management for the sake of someone else'.[82] Stewardship referred to the nameless 'future generations', which should become a prominent point of reference in the sustainability discourse of the late 1980s.[83] Stewardship also implied a quasi-divine power that had trusted the earth to human administration. Stewardship led back to the fundamental question of the position of humans in nature. The modest self-understanding of humans as the wardens of the earth on behalf of a third party itself took on heroic qualities. Marshall McLuhan had anticipated this self-image of active human responsibility for and contribution to the planet's fate in 1964: 'There are no passengers on Spaceship Earth. We are all crew'.[84]

In 1972, the study *A Blueprint for Survival* projected a rigorous goal- and future-oriented action programme for humankind. Published by the editors of the British magazine the *Ecologist*, founded in 1970, the *Blueprint* demanded to confront the 'evidence of the total ecological crisis with which mankind is faced' by facing the facts, as editor Tom Stacey appealed in his foreword.[85] The authors called for overcoming the collective bemoaning of environmental deprivation in the retrograde and futile hope to restore nature. Rather, they claimed to take a rational approach to the present for the sake of the future by creating an extensive register of accumulated and systematized pressures, depletions and corrosions of human society and the environment. While *Planet in Peril* had worked mainly with text and photography, tending to individualize environmental damage to certain locations in remote regions, the *Blueprint* synthesized and collectivized destructions, aggregating and redistributing responsibilities on a global scale. In the sum of sober tables and diagrams of worldwide resource bases, and in the accumulated statistical data of calculable rates of production and consumption of nature, the global systems gained shape and became the new objects of extensive strategic measures and detailed synchronizing schedules. A fully-fledged step-by-step plan was to secure the goal of stabilizing the 'ecosystems' and the 'social systems' in the twenty-first century.

Stock Maintenance: System Earth

The *Blueprint* transformed the image of the delicate Whole Earth into a giant expert system of which the blueprint, the elaborate guidebook, was just being written. The expectations to draw up the detailed plan and programme of action for human survival were directed towards the sciences and engineering and to their methods of rational, problem-oriented and process-minded analysis and planning. The notion of being able to control the earth system in its totality was fostered by the distanced view of the planet from the outside. The new image of the earth as a spaceship connected with the vision of a sustainable planetary management. Spaceship Earth went into coalition with System Earth.[86]

According to Boulding's programmatic view on Spaceship Earth, a strict economy and technology of cyclic material energy and information flows were waiting to be developed to ensure stock maintenance. Systems scientists literally took the standpoint of the scientist-astronaut viewing the earth from space, while using the image metaphorically to outline a new abstract concept of the earth: 'We can begin a systems view of the earth through the macroscope of the astronaut high above the earth', claimed the American systems ecologist Howard Odum in 1971. 'From an orbiting satellite, the earth's living zone appears to be very simple. The thin water- and air-bathed shell covering the earth – the biosphere – is bounded on the inside by dense solids and on the outside by the near vacuum of outer space.'[87]

The ecosystems sciences mobilized a utilitarian approach to nature that relates back to natural philosophers like Francis Bacon and Carl von Linné. Referring to an earlier classification by the environmental historian Donald Worster, Andrew Jamison has characterized this tradition as 'imperialist'. Supplementing Worster's two historical views of humans' place in nature, the imperial or dominating approach and the Arcadian or romantic approach, Jamison contrasts three traditions of the environmental movement that became active again in the 1960s: the Arcadian tradition inspired by the work of Rachel Carson on environmental contamination, the human-ecological tradition initiated by Paul Ehrlich's work on the relation of population and resources, and the imperial tradition of scientific ecology of which the brothers Howard and Eugene Odum were major proponents.[88]

Ecosystems research conceived of nature as an economically operated system made of single parts that worked together functionally and efficiently. This vision was 'imperialist' since it conceptualized nature as a machine run by a human operator. Moreover, in analogy to the effort of engineering nature, of framing nature entirely in mechanical and systems terms, systems sciences introduced the project of 'ecologizing society', of framing society entirely in the general systems terminology of energy, matter and information. The Odums in particular encouraged this strategy as a political mission.[89] The capability of managing the global social-natural earth system was directly – and in the opinion of social systems engineers also consistently and inevitably – derived from the systemic models that simulated the laws and causalities that natural scientists assumed operative in nature.

In ecosystems science, the concept of the biosphere was supplemented and filled by a number of different neologisms that were used more or less synonymously, like 'global environment', 'ecosphere' and 'ecosystem'.[90] The biosphere entailed a spatial delimitation of the earth that allowed for the estimation of the sum total of its machine-like parts and suggested unprecedented potentials for future human intervention and control. But while the concept of the biosphere had been tied primarily to a geologically and biologically defined space of measurable dimensions and qualities, the 'ecosystem' was based on an analytical mathematical framework. The limits, the norms and standards, as well as the scale of the system had to be selected according to choices based on analytical questions and preferences; a system rested on relative grounds. The ecosystem possessed no geographical boundaries apart from the point of view of the scientist-observer. Nevertheless, the new world and earth systems were represented as referring to a bounded earthly space with geographical dimensions. This was a common connection that is exemplified in one of the famous publications of the time: *The Limits to Growth* was the Club of Rome's report on 'the Predicament of Mankind' from 1972 that turned into a bestseller.[91]

The Limits to Growth superposed geographical and systemic limits so as to be physically and scientifically congruent. The study established equivalence between the abstract calculations of resource economics and the perceptions of geospatial limitation that had become familiar in the Whole Earth images. The MIT study group designed a World Model, which created physical and scientific equivalence in the language used, in the computations it was based on, and also in its visual expressions, coupling graphs and diagrams. The World Model was represented as a flow diagram in formal system dynamics notation. Different symbols represented different states of different variables: rectangles represented physical quantities measured directly, valves represented rates influencing the levels of quantities, circles represented auxiliary variables influencing the rates. Sections in rectangles signified time delays. Arrows visualized flows, solid arrows pointed out material flows and broken arrows indicated causal relationships. Clouds represented sources or sinks that were deemed insignificant for the model behaviour.[92]

The computer-based mathematical and numerical routines used for modelling the world in its totality were devised according to prototypes *World1* and *World2*, developed by Jay Forrester and introduced in Forrester's book on *World Dynamics* from 1971.[93] *World3* was the name of the advanced computer model the study applied to create different scenarios of economic and ecological development. The model encompassed the world by means of five highly aggregated major social-natural indicators: 'resources', 'population', 'food per capita', 'industrial output per capita' and 'pollution'.[94] By arithmetically computing the progress of these variables over time, the future development of the world was modelled and plotted in curve diagrams. The 'standard run' was based on the assumption of unchanged major physical, economic and social relations up to the year 2100, and it resulted in a threatening scenario of rapid growth and 'collapse'.[95]

The rising curves of population, production and food supply depicted the massive industrial growth since the turn of the twentieth century. They reached a peak in the middle of the twenty-first century before they dropped off as swiftly and dramatically as the overarching resources curve. Each variable was plotted on a different vertical scale, while these scales, as the authors note, were omitted altogether; the researchers deliberately withdrew the scales and kept the horizontal timescale vague 'to emphasize the general behavior modes of these computer outputs, not the numerical values, which are only approximately known'.[96] And yet, offering more scientific detail than Dasmann's photographs of the *Planet in Peril* and less open-endedness than the registers and timetables of the *Blueprint*, this single computer plot mediated the message of alarm and of acute pressure for action. The apprehension was contained in the motion of the curves: 'Is the future of the world system bound to be growth and then collapse into a dismal depleted existence?'[97]

In its conceptual and arithmetic abstraction and simplification, the study suggested completeness and integrity. It contained a statement about the entire 'world system' and its fate. In simplifying a complicated situation of relations in nature and society by integrating selected components while leaving out others, and by coupling components based on direct functional relations, the study presented an image of the world that not only made sense but that also became a convincing figure of how the world worked: as a huge contraption organized by way of functional interdependence. Through the projection of abstract systemic limits on the geographic limits of the earth, the limits of nature were made scientifically accessible and determinable to the systems operator. The study carried forth that it was possible to speak not plaintively but knowingly about the earth.

The study *The Limits to Growth* exhibits Schellnhuber's vision of systemic closure and of complete exposure of the earth in its embryonic stage. The physical earth had shrunken rapidly with the rise of a global physics that Alexander von Humboldt could not have imagined in closer detail. By mid-century, the earth sciences could draw on long-term, earth-spanning observations, on new infrastructures of computer-based data processing and on growing networks to circulate information on a global scale. In the second half of the twentieth century, the systems sciences emerged as a new 'interdiscipline'. The interdiscipline, a notion emphasized by Geoffrey Bowker, epistemologically recreated "'the world" as an ecological and physical unity', as Paul Edwards phrased it.[98] New in the world models, the computer programs to model the 'world dynamics' for *The Limits to Growth*, is the practice to treat the earth systems as closed systems. The notion of systemic closure enabled mathematical modelling and forecasting, but it constrained cultural difference and political negotiation. Closure favoured the notion of the earth as a common interior space, uniform and regular. And it facilitated the idea of rerunning and eventually recreating the world from scratch. Much of the uneasiness of the Cold War era falls between this ambivalence of the earth's future, a future made predictable and put at stake at the same time.

The world seemed, in principle, accessible and engineerable like the spaceship that had become its prominent metaphor. Throughout, the study applied the theories and terminology of feedbacks and behaviour modes, of control and regulation from systems dynamics and cybernetics. The earth was conceptualized as a closed system consisting of positive and negative feedback loops of social and natural processes alike, in which humans could intervene at will. From the 'standard' expectation, possible alternate development scenarios were derived by mathematically modelling the central, coupled variables under different boundary conditions. The study's 'stabilized' world models displayed different desirable 'states of global equilibrium', depending on the variables that were taken into account and governed through measures of growth regulation; these were, for instance, birth control, life cycle improvement, resource recycling, pollution control and capital stabilization.[99] The desired equilibrium state expressed the prevailing phantasm of controlling the parameters of all world processes to achieve ecological homeostasis.

Intervention: Towards a Planetary Machinery

'The future is not to be predicted, it is to be planned'.[100] The American designer Medard Gabel voiced this ambitious claim in 1974 in his book on *Energy Strategies for Spaceship Earth*. Gabel was a disciple of the American architect and designer Richard Buckminster Fuller, and with Fuller he was a co-founder of the World Game Institute in 1972. The World Game was a large-scale earth laboratory set up as a participatory game to enlist people to develop strategies for 'how to make the world work'.[101] Gabel's motto concerned not future scenarios; rather, the research group aimed to create a concrete plan. To plan 'the future' in the singular required a linear consciousness of time, a philosophy of progress and an idea of perfection which in previous centuries had not been possible to achieve during one's lifetime; if at all, fulfilment was accomplished in the afterlife.

When *The Limits to Growth* was published, however, the time when future had no name but held only vague hopes, at most prospects, had passed. Nothing could be gained by claiming to hold good hopes; it became important to nail down solid expectations. In the discourse of forecasting and planning the future, expert groups of scientists and politicians aimed at a future in the singular that would be ready for planning if only the relevant mechanisms were studied and all data necessary were accessible and accumulated. The mathematician Norbert Wiener, the founder of cybernetics, had propagated this universalistic vision based on a strict polarity of order and disarray, control and chaos. Wiener's cybernetics, the rising science of automatic control engineering, had been a genuine war endeavour. Wiener sought to transform the contingent and unpredictable world into reproducible algorithms, formal operations and routines. With the help of mathematical and technical methods of communication and information, he recreated the world in terms of feedback circuits.[102] Moving beyond modelling and simulating the present, cyberneticists aimed to reduce possible futures by employing their 'teleological' machines, providing for the one desired future to manifest itself. The future that had long been principally unknown and open-ended became an object of technical regulation and control.

The world models of the study *The Limits to Growth* perhaps provided the best-known examples of large-scale models based on a theory of systems that encompassed both natural and social phenomena.[103] The authors insisted that they would never dare to predict the future, since they operated on a merely 'intermediate ground of knowledge'. They stated that theirs, rather, was an attempt of unfolding probable futures by roughly outlining pictures of development.[104] Nevertheless, the fundamental assumption held that if the data were improved and information was complete, the models would change from a possibility to a prediction. The question of whether the predictions of *The Limits to Growth* and similar studies were accurate, or whether the prognostic power, the data collections and the computing methods yet to come would prove more precise is beside the point. A more essential criticism was being voiced already at the time of the study against the unbroken optimism of principally being able to model and to steer the development of the world. Open to question was and is the claim to grasp natural and cultural developments in total and to be able to direct their future behaviour.[105]

Science studies scholar Paul Edwards has cautioned against accepting the idea of pure data and pure models. Ever-more complex models and ever-increasing computing power, he argues, will not be able to overcome the constraints of scientific method and the corresponding irreducible epistemological uncertainties.[106] Encapsulating complicated realities in global algorithms and enrolling large-scale entities into simple equations literally disrupt social, political and geographical dimensions. Likewise, Peter Taylor and Frederick Buttel have criticized the expert cultures and their aggregating practices, 1970s style, as a moral-technocratic alliance that bypassed the political to define what counted as 'environmental problem' and how to solve it.[107] After all, the cybernetic concept of environmental governance was based on a problematic analogy of technological and political authority. The *governor*, from the Latin *gubernator*, is a translation of *kybernetes*, the Greek term for 'helmsman'.[108] Thus, *kybernetes* refers us back to Plato's ideal state as a ship on high seas steered with a firm hand and subjected to a strict hierarchical organization.

The Biospheric Life Support System

> One of the modern tools of high intellectual advantage is the development of what is called general systems theory. Employing it we begin to think of the largest and most comprehensive systems, and try to do so scientifically. We start by inventorying all the important, known variables that are operative in the problem. But if we don't really know how big 'big' is, we may not start big enough, and are thus likely to leave unknown, but critical, variables outside the system which will continue to plague us.[109]

'"How big can we think?"'[110] In answering his own question, Buckminster Fuller recapitulated some of the main lines of reasoning in the field of world system dynamics that I have roughly sketched in this chapter. When Fuller published his *Operating Manual for Spaceship Earth* in 1969, he took the metaphor of Spaceship Earth to a literal use. 'We are all astronauts', Fuller argued, and he readily conceived of humankind as adapted to travelling though space at sixty thousand miles an hour on its very big planetary spacecraft.[111] Moreover, he states, 'Spaceship Earth was so extraordinarily well invented and designed that to our knowledge humans have been on board it for two million years'; humans have even been 'able to keep life regenerating on board'.[112] In contrast to Ward's and to Boulding's references to Spaceship Earth, Fuller stressed neither the need for political unity nor for a future steady-state economy. He focused on the total operating scheme of the earth as a technological system: 'We have not been seeing our Spaceship Earth as an integrally-designed machine which to be persistently successful must be comprehended and serviced in total'.[113] Fuller stretched the metaphor of the spaceship provocatively. Since 'no instruction book came with it', humankind was confronted with the challenge of learning on its own 'how to operate and maintain Spaceship Earth and its complex life-supporting and regenerating systems'.[114]

Fuller's picture of Spaceship Earth is in line with the notion of the world system in its state of dynamic equilibrium that accounts for the human environment through the integration of regenerative functions, or 'life support'. The imagery of life support was taken from space science and became codified in the scientific definition of the biosphere as an '*integrated living and life-supporting system*'.[115] This assessment of the biosphere as a planetary supply and regenerative system adapted the idea of the spherical living space first suggested by Sueß. It took up the concept of earthly living space as a 'living organism'[116] that was promoted by Vernadsky and later by Lovelock. Moreover, the biospheric life support system incorporated Boulding's thermodynamic view of the earth as a dynamic system of material and energy flows. Last but not least, the life support system integrated the expectation of being able to develop biospheric technologies to sustain the system. Life support accommodated arcane images of a holistic nature, but transformed them into the view of a machine with automatic control.

Howard Odum was among the first to discuss the use of ecosystems science for life support in space travel.[117] The Apollo 13 accident in April 1970 demonstrated some fundamental differences between artificial life-support systems and the life-support system of the biosphere. The spacecraft was essentially a storage vehicle that carried sufficient food, water and oxygen, and accumulated human wastes for later disposal. Supplies could not be replenished, wastes not recycled. In contrast, the biosphere was conceptualized as a regenerative system, controlled by complex self-regulatory or homoeostatic mechanisms.[118] Odum, however, perceived these differences not as *epistemological* but as *technological* in character. After the Apollo 13 incident, the analogy of Spaceship Earth to him seemed even more appropriate, precisely because the earth, as Joel Hagen clarifies, 'had no safe haven to which it could return. Therefore, damaging the life-support systems of the biosphere was courting disaster'.[119]

The *Scientific American* special issue of 1970 on the Biosphere introduced earlier in this chapter was but one remarkable example of the globalization of the ecosystems perspective in the later twentieth century. The articles conceptualized the biosphere as the largest dynamic ecosystem of the earth and explored it in terms of a cyclical system and ecological metabolism. Each article covered one of the 'major cycles' of the biospheric life support system: energy, water, carbon, oxygen, nitrogen and minerals, as well as the human production of food, energy and materials that contributed to the planetary metabolism. The authors understood the large ecosystems of the earth: forest, ocean, marshland, tundra and desert, as a closed system that allowed the exchange of energy but not of matter with its surroundings. The energy source was the sun. Like an engine it powered and maintained all 'living systems' and 'life-support systems'.[120]

The differences between the life-supporting systems of a spacecraft and of the earth's biosphere had become marginal. Biologist Paul Ehrlich and political scientist Richard Harriman organized their 1971 book *How to Be a Survivor: A Plan to Save Spaceship Earth* entirely around the metaphor of Spaceship Earth, with chapter titles such as 'The Size of the Crew' and 'The Control Systems', along with a chapter describing the new 'Spacemen' culture that would have to develop.[121] While at first sight these discursive shifts may seem small, the following chapters will demonstrate how powerfully Spaceship Earth and its life support system reorganized the traditional inclusive images of the 'good ship Earth'[122] on the verge of sinking.

4 STORAGE: THE LIFEBOATS OF HUMAN ECOLOGY

A Clever Montage

In a two-minute introductory sequence the movie *Soylent Green* outlines the history of the twentieth century. Consisting of pictures and music, the sequence starts out with long shots documenting the optimistic times of the rising bourgeoisie, growing industrialization and increasing domination of nature at the turn of the century, symbolized by quaint patriarchal family circles and groups of proud mountaineers as well as by the first fragile cars and primitive airplanes. A slow waltz accompanies this buoyant but still moderate progression into modernity. As the century proceeds the historical pictures are progressively replaced by images of modern industrial settings and of the output of mass production; we see smokestacks, workers toiling away and arrays of identical cars leaving the assembly line. As the music picks up speed and changes to a quick beat, the succession of pictures accelerates as well; the screen splits as images double and then multiply, changing with increasing frequency. Also, what we see changes: in frantic succession more and more people appear on the screen. We see people moving in busy urban crowds in the 1950s and 1960s. The images are in colour now, denoting post-war prosperity and the affluent lifestyle in the Western world. From within this frenetic rhythm of pictures and music, the effects of industrial growth and economic wealth appear in the form of dead trees, waste and of people wearing facemasks in the thick smog of megacities. The music returns to the slow waltz from the beginning, though it now strikes a rather gloomy tone. Also the images slow down, and the viewers note that they have again moved on in time: Now we see polluted wastelands, destroyed forests and barren industrial sites. The sequence 'comes to a grinding halt' with the sight of the thickly polluted cityscape of New York City. The year is 2022. The population is forty million.[1]

Thus closes the introduction to *Soylent Green*, a troubling dystopia produced in the United States in 1972 and released in 1973. The movie presents a vivid imagination of how the abuse of natural resources, the growth of population, industrial contamination and the greenhouse effect would destroy urban environments in the near future. Heat and smog make life almost unbearable. Fresh food is hardly available. Congestion, poverty, hunger and corruption dominate the city. A huge police force keeps the masses in control. A single company, the Soylent Corporation, controls food production and distribution. Fresh vegetables, fruit and meat are a luxury of the rich. The masses are fed with synthetic nutrients based on proteins from soybeans and ocean plankton. Their food rations come in small, tasteless bits and pieces: Soylent Yellow, Soylent Red and Soylent Green, sold on Tuesdays to the angry, starving crowd.

The screenplay of *Soylent Green* was based on a novel by Harry Harrison titled *Make Room! Make Room!* that appeared in 1966.[2] The book's title carries a dual meaning. It signifies the unbearable constrictions of an overcrowded world as well as the merciless practices of riot control by means of the local police forces that push through the throng with huge power shovels and clear people away with garbage trucks. What is commonly classified as science fiction was, according to director Richard Fleischer, neither about science nor about fiction and fantasy. Rather, Fleischer's scenario was meant to be poignantly 'prophetic' by imagining the situation of too many people who share too little space. The movie's theme of extreme population increase, environmental degradation, scarcity, mass uprising and mass mortality had also been the topic of several outstanding works of the era. Among the best known are Fairfield Osborn's *Our Plundered Planet*, published in 1948, *The Limits of the Earth* in 1953, and *Our Crowded Planet* in 1962 – not to forget Karl Sax's *Standing Room Only* in 1955, Paul Ehrlich's *The Population Bomb* in 1968 and Georg Borgstrom's *Too Many* in 1969.[3]

Soylent Green linked two pressing discourses of its time: 'overpopulation' and 'overpollution'. Fleischer's introductory montage sets up this link. The impressive tableau follows the movements of exponential growth and stagnation, or 'population explosion' and 'ecocide', to use two widespread terms of the time. In its visual and acoustic structure the 'cleverly devised' sequence exactly matches the Biological Growth Curve to which population ecologists referred to describe the development of a population over time. This S-shaped curve related population size and time according to a natural law of population growth. The constraining factor in population growth patterns, biologists explained, consisted of the confinement of most populations to a strictly limited area. Populations would develop according to such limitations; their growth would be exponential up to a point of inflection when environmental feedback would cut in, and subsequently, progressive deceleration and decline would occur.

The underlying *mathematical* model had been proposed in the nineteenth century and was re-established in the 1920s, derived from observing self-contained fruit fly populations bred within sealed glass jars. Nevertheless, the model was soon held valid to describe human population development as well. The first half of the curve had been familiar since the early 1800s, when the Reverend Thomas Robert Malthus published his thoughts on the 'Principle of Population'.[4] Malthus had proposed straightforward numerical rules to forecast the disparate development of population and food supply. According to these rules, the number of people increased geometrically, while food supply tended to increase arithmetically. In the Environmental Age scientists readdressed the Malthusian predicament and anxiously repeated the long-standing question of 'How Many People Can the World Support?'[5]

The *empirical* world population growth curve, which involved plotting population figures from ancient times up to the present and extrapolating population into the year 2000, explains why the contemporaries deemed their situation so unique to history. More disturbing than absolute population numbers was the 'nature' of growth. Growth was considered to be of an exponential nature. Within less than a century world population had doubled to roughly three billion people and it increased at a rate that led to anticipate another doubling within merely one generation. The 'doubling time', the time span in which world population would increase to twice its number, had shortened to a mere thirty-seven years.[6]

Population growth threatened to surpass the economic expansion of the post-war period. The achieved standards of living in the West and subsequently in the entire world were projected to collapse within two decades. 'In just two or three years it became possible to question growth, to suggest that DNA was greater than GNP', David Brower put it, the managing director of the American environmental organization Sierra Club.[7] 'Unprecedented growth' became a common term in scholarly treatments of the 'population problem': (Western) society considered itself positioned at the *end* of an exponential process of growth at which the limits would appear very suddenly. Human ecologist Paul Ehrlich nailed down the issue at stake: 'Clearly a long history of exponential growth does not imply a long future'.[8]

The Biopolitical Nature of Statistical Aggregates

Population accounts are intriguing tales.[9] First and foremost, the impressions of increasing or decreasing human numbers conceal the uncertainties involved in the estimates of the size and development of a nation's or the world's population. The precision with which numbers are linked to specific moments in time results from statistical computations that aggregate different quantitative demographical averages into a superordinate assemblage. In aggregate, population may then be accounted for to the exact year and day, hour and minute. Second, statistics lends the abstract unit of population the epistemic status of a concrete life form with a past and a future development, to be reasoned with and operated on. This natural body generates assurance and certainty as well as doubts and fears, depending on when, how and by whom it is assembled, visualized and referred to. Population might be treasured in its aggregate form, statistically averaging across individual characteristics like age and origin, but it may raise anxieties when analysed into differentially growing groups. Thus, thirdly, practices of pulling together human beings into enveloping populations are powerful political acts that highlight certain aspects of individuals and groups over others. Deciding on who or what counts, and how, mirrors and reinforces political and social values. The seemingly neutral and objective, even compelling practices of assembling and aggregating humans not only define social categories but redefine the order of the social.

This chapter explores the question of how humans in the Environmental Age were statistically aggregated into populations and how these populations were discriminated into valuable or dispensable lives in accordance with prevailing economic and ecological regimes. I will focus on the history of 'world population' as a historically recent statistical collective.[10] World population emerged from the growing scientific potentials to collect earth-spanning data in the late nineteenth and early twentieth centuries. World population took concrete shape in the second half of the twentieth century within the systems sciences. Based on long-term observations, on computer-based data processing, and on the growing scientific networks that circulated information on a global scale, population was assessed, modelled and forecasted. As a 'variable analogous to capital, labor, technology, or infrastructure in a "world system"', population thrived in the developmental and environmental discourses of the 1960s and 1970s.[11] The relentless growth of world population was met not with appreciation but with mounting apprehension. In the Environmental Age, when limits to growth were not only forecasted but also numerically predicted, the discourse of a global 'population crisis' drew on ecological evidence of a world suffocating from overpopulation.[12]

In the 1960s demographers counted about three billion human beings populating the earth, less than half of present estimates. However, when analysing the discourse of overpopulation in the Environmental Age, comparing absolute numbers in hindsight to show far-sightedness or flaws in the historical conceptions is misleading. Instead I will explore the perceptions of the 'nature' of population growth around 1970. I will trace the dynamics of the population debate and the scientific practices of monitoring, evaluating and disciplining unruly population on a planet that had been framed in the terms of a spaceship. The new field of human ecology is particularly rewarding with regard to Spaceship Earth since proponents conceptualized longstanding questions of human living space as a conflict of global ecological 'capacity'. World population was calculated as a function of the upper limit of the closed earth system, the earth's 'carrying capacity'.

In line with models of ecological balance, carrying capacity accounted for the maximum number of human lives that could be sustained within the limits of planet earth. In the 1960s, the concept of an ultimate earthly capacity seeped from biology into human ecology and demography, and spread to the vocabulary of UN officials, political decision-makers and economic advisers on a global scale.[13] As pointed out in Chapter 1 of this book, Michel Foucault had noticed in the same decade 'that the anxiety of our era has to do fundamentally with space'.[14] Identifying his epoch as one of simultaneity and juxtaposition, in which space took the form of relations among sites, he conceptualized the 'problem of the human site or living space' as the problem of knowing about the 'relations of propinquity' and the 'type of storage, circulation, marking, and classification of human elements' to be adopted in a particular situation to achieve a given end.[15] As a consequence Foucault foresaw the emergence of a 'human topography', a discipline accounting for the siting and placement of human elements according to a combination of geographic, technoscientific and economic expertise.[16] Extending Foucault's projection I argue that the notion and assessment of a population crisis reformulated the theme of contested living space along the lines and divides of bio-economical eligibility commensurate to a limited ecological capacity.[17] The concept of carrying capacity I understand as another expression of the spaceship that the earth had been turned into.

I am interested in how the concept of carrying capacity was used to outline and define an ecologically sound global community. Furthermore, I ask how carrying capacity justified selection processes among people and entire nations according to their ecological function and fitness, selection processes that determined who would be able to stay and travel on the earthly spaceship and who would have to disembark. The scientific validation of a limited ecological carrying capacity was established in numerical and mathematical equations as well as in visual images and graphs. I will scrutinize the example of the 'biological law of population growth' and its smooth S-shaped curve describing population size over time. This law statistically linked population growth and environmental degradation on the global scale. My aim is to reconstruct the development of the law from its fabrication in population biology in the early twentieth century to its featuring in ecological development studies in the 1970s.

After a brief look at the statistical means for providing evidence of populations in numerical terms, I will examine how the mathematical model of the biological law of growth was derived experimentally in the 1920s and how the claim took hold that the law supplied evidence of the self-limiting growth of *any* self-contained population. The shift from biological 'breeding groups' to human populations solicits further exploration of the status and position of ecologists in setting the terms of the population discourse in the Environmental Era. I will pursue the standards set for the 'normal' development of human populations and for its quantitative deviation, 'overpopulation'. The final sections will focus on the moral economy inherent in the law of population growth and its associated concept of carrying capacity. As the growth law assigned populations to the realm of the life sciences, it gave preference to numerical and bio-economical over social and political explanations and solutions in areas of environmental and developmental concern. For human ecologists, the growth law warranted the aim of bringing about an optimum population and supported demands for strong measures of population control, taking human lives into account as manageable, marketable or disposable.

The Life of P in Numbers, Words and Images

'Let P = population'. With this equation, historian Barbara Duden brings to mind that population has never been a natural given. Populations came to life in the late eighteenth century as aggregates of statistical data 'used to generate the semblance of a referent' which would then be addressed as the real object of development policy.[18] Within the statistical framework of reasoning, Duden claims, '"P" acquires a life of its own'. Malthus's bold claims of 1798, based on sparse data, helped the onset of 'social physics', the study of the mechanics of the new national societies composed of individual elements, to be dealt with in mathematical terms. With the rise of population statistics in the nineteenth century, constant patterns were imposed on the previously fleeting and ever-changing forms of life.[19] Populations were turned into natural subjects which 'grow, consume, pollute, need, demand, are entitled', Duden observes. At the same time, she remarks, populations became natural 'objects that can be acted upon, controlled, developed, [and] limited'.[20]

Historian of science Sharon Kingsland has shown in her study *Modeling Nature* that political economists like Malthus, empiricist social scientists and statisticians like Auguste Comte and Adolphe Quételet, as well as natural philosophers like Charles Darwin were fascinated by 'the quantity of life' and the ways in which this quantity was held 'in check'.[21] In this transition of life to a state that could be quantified and modelled, 'a new language came into being', Duden adds, 'created to observe people in quantitative contexts'.[22] Demographic accounting practices, formal mathematics and statistics were constitutive of the new human collectives acquiring shape. Representations of populations as self-contained entities in terms of population dynamics, of population explosion or, in the late twentieth century, of population screenings, were language acts trimming human beings to fit a single common denominator for the sake of accountability.[23]

Biologist Karl Sax, professor of botany at Harvard University, proposed in 1955 that with the present rate of population growth 'in 600 years the entire earth would provide only one square yard of land per person' – there would be 'Standing Room Only'.[24] Equally alarming was the French Riddle of the 'global lily pond'. The literary image was widely applied in the 1970s to convey the threat of exponential population growth. In 1972 the study *The Limits to Growth* employed this analogy of human beings and water lilies. If, so the riddle went, a lily pond of single lily leaves whose number doubles each day is completely full on the thirtieth day, when is it half full? The answer was: on the twenty-ninth day. Lester Brown, president of the World Watch Institute, titled his book from 1978 accordingly, *The Twenty-Ninth Day*, and warned that the global lily pond may already be half filled.[25]

Record-keeping techniques and statistical tools accomplish homogeneity, comparability and, eventually, the mathematical commutability of human elements which were never the same or even similar. Human beings could be accounted for as elements in larger plots and pictures, outlining spatial arrangements and characteristic movements in time. Statistical accounts of impending human congestion combined the Malthusian principle of exponential growth with the suggestive images of asymptotic population saturation, of the regression and standstill that the biological law of population growth and its S-shaped curve conveyed. Moreover, demography in the same breath 'normalized' the created population by rearranging and aligning individuals according to statistical averages that would then materialize as social norms.[26]

Numbers know, images show. Perhaps more powerful still than scientific tables and equations collecting quantitative knowledge are graphs and images, which act as normalizing devices that transform cultural narratives into normative visual representations.[27] A simple table with two columns and a couple of lines effectively cut down the characters needed to explain Malthusian geometric or exponential population growth. The table's sequence again was weak compared to a chart plotted to scale, as Jay Forrester admitted, himself a master of computation and plotting.[28] By the 1970s, neo-Malthusian exponential growth curves were distributed widely, enabling an individual to gauge the situation of its collective present, of its past and of its imminent future at one glance. Whereas a table simply tabulated numbers, the exponential growth curve played to its strength in the choice of vertical scales, thus making 'growth appear so steep and sudden'.[29]

The Law of Population Growth

Studies of how Darwinian selective pressures and Malthusian checks acted on populations opened up a field in the 1920s that within a decade would be recognized as population ecology. Population ecology at first was not directed at human populations but operated mainly in the field of entomology. The study of small insect populations formed a link between evolutionary biology and ecology. The new discipline of population ecology provided knowledge of the regularity and comparability of populations and supplied a mathematical description of their nature. 'The questions involved numbers, and the answers seemed to lie in the direction of mathematics', Kingsland remarks in her chapter on the early history of population ecology. 'Before too long', she continues, 'the mathematical answers began to appear'.[30]

Population ecology experimented on single-species aggregations and on population cycles. Biologist Raymond Pearl at Johns Hopkins University combined a mathematical point of view with an experimental style and practice. Studying *drosophila*, fruit flies, Pearl devised a programme for a 'comparative demography', the systematic attempt to apply demographic techniques to animal populations.[31] In his laboratory Pearl used bottles of specific volume, in which he placed measured quantities of nutrition. Simple half-pint milk bottles, for instance, would initially contain a small number of animals. In such controlled settings fruit fly populations would increase exponentially up to a point when population size would outweigh the amount of food remaining. Fruit fly numbers would decrease and approach an upper limit that was defined by the restrictions of the bottle.[32]

A concept of constraint had long been familiar to biologists. It can be traced back to the Law of the Minimum that the German chemist Justus von Liebig had formulated in the mid-nineteenth century for agricultural chemistry.[33] The law stated that the yield in agriculture – for example, from a farmed field – varied according to the field's size and its regenerative powers. Liebig held that the law of the minimum was valid for all nutrients: 'Every field contains a maximum of one or more nutrients and a minimum of one or more other nutrients. The yield will be in proportion to the minimum ... it governs and determines the amount or the extent of the yield'.[34] Liebig spoke of a 'natural law' that he called the Law of Replacement. In its generalized form, 'Liebig's Law' predicted that populations of any species would be constrained by whatever resource needed for survival was in shortest supply.

Pearl opened up a new academic field to exemplify Liebig's law of the minimum. As Liebig had analysed soil into single elements and investigated their composition and their productivity to estimate the sustainable yield, Pearl analysed animal populations into their single parts to determine their maximum yield or 'reproductivity' in relation to the available provisions. Biostatistics became his method to study the new 'biology of groups'. In a systematic effort to apply demographic techniques to animal populations, he borrowed from social statistics the established tools of life tables, birth rates, death rates and life expectancies. When official assessments of the First World War brought the Malthusian predicament of population and food supply back to the national agenda, Pearl set out to study human population growth rates together with the mathematician Lowell J. Reed. The sought degree of mathematization he found in the accumulated data body of social statistics, on fertility, growth, disease and mortality.[35]

Pearl's 'search for quantitative measures and correlations under different experimental conditions' aimed at establishing a *law* of human population growth, expressed as a mathematical equation which would conform to experimental observations and indicate future trends reasonably accurately.[36] He became a convert to the statistical view of nature that Karl Pearson, the British founder of statistics and academic father of Francis Galton, the founder of eugenics, had advocated.[37] Pearl believed in unbiased, consistent and comprehensive scientific method to formulate scientific 'laws'. Deriving sense from the order of facts, and from the relationships between isolated phenomena, science would represent the fundamental 'laws of nature'. A natural law, according to Pearl, emerged from a rational process of description, theorization, generalization and explanation.[38]

Pearl and Reed presented their population equation in 1920. They adopted the term 'logistic' for their smooth, S-shaped curve towards a stable upper limit. The term 'logistic' was a reference to the Belgian Pierre-François Verhulst, who had described population growth in 1845 and had called his curve of population growth over time 'logistique'.[39] Following Kingsland, the curve eventually 'contributed to population ecology one of its simplest mathematical models'.[40] The aesthetics of the growth model were compelling; practically, the curve imposed on the observer a prescribed reality. Moreover, to quote Pearl, the model demonstrated that 'plainly all growth, including that of population, is fundamentally a biological matter'. Whether humans or yeast cells, the course of all populations was determined by the same natural law and consequently had to be subjected to the authority of the life sciences.[41]

Nonetheless, Pearl's basic assumptions raised controversy among his colleagues. On the one hand, the symmetry of the curve seemed blatantly idealistic. It implied that the forces in the first half were exactly as strong as in the second half, with the point of inflection exactly halfway through the curve. On the other hand, a more general equation that would free the curve from its restrictive symmetry would also have entailed that the curve would fit virtually any data. In this case the curve could not have been considered a calculating device with prognostic value and hardly a natural law in the strong sense. Nevertheless, Pearl and Reed held on to their smooth mathematical tool. They defied the empirical clutter they had to accommodate and enforced law and order among their experimental populations. Following Kingsland, Pearl and Reed 'had assumed that which had to be proved, yet they presented their curve as being empirically fitted, not logically derived'.[42]

Pearl published his main work on *The Biology of Population Growth* in 1925. The book illustrates how Pearl accumulated the rather sparse data on human populations from different countries that were available at the time, how he arranged the data along the curve line and extrapolated the data to fit his logistic world population growth curve. 'Pearl had come full circle: having first *assumed* logistic growth in trying to find a curve to fit his initial data, he now believed that the empirical evidence proved the truth of the logistic "law", even though large parts of the curves were extrapolated', Kingsland comments.[43] But Pearl accomplished more. Studying animal populations to develop the model and then applying it to human population development, Pearl succeeded in framing and articulating human collectives in terms of biological 'breeding groups', reducing human lives to one single characteristic: biological reproduction, indicated by group size. Pearl assigned to world population the epistemic status of a simple biological life form with a stable course of development and a predictive lifecycle. This shift allowed him to merge the growth model with extrapolated human developmental trends and to spark the idea of human populations as entities that would behave according to natural principles.

Behaviour Modes and Survivorship Curves

Pearl co-organized the first World Population Conference in Geneva in 1927 where he was introduced as one of 'the world's leading experts on the phases of the population problems; biological, social, economic, medical, statistical, and political'.[44] He presented his law of population growth and discussed possible means of influencing the terms of the law and controlling the gradient of the curve. The Geneva conference can be understood as an indicator of the focal shift from the national economies of the nineteenth century to an economic prosperity governed by international relations after World War I.[45] Only in the mid-twentieth century, however, did demographic assessments and forecasts enter the discourse of development. Duden distinguishes three periods of population discourse since 1950, each through statistical aggregation obliterating 'ever more thoroughly the consideration of real people'.[46] In policy statements of the 1950s population still figured as an equivalent to concrete social collectivity, and overpopulation just appeared as a threatening image on the horizon.[47] During the 1960s, population policies proliferated and ideas of population control moved to the vanguard of state politics, owing as much to the awareness of growing global political interdependence as to rising concerns about environmental depletion and pollution.

At the beginning of the 1970s, books and articles appeared in great numbers on the 'unprecedented increase in human numbers which has gone unchecked since the Industrial Revolution', on 'uncontrolled growth' and 'overcrowding', and on the need for governmental regulations to enforce birth control.[48] 'Few issues in the world have undergone such a rapid shift in public attitudes and government policies over the last decade as the problems of population growth and fertility control', George Herbert Walker Bush, then US Representative to the United Nations, proclaimed in 1973.[49] Scientists, ecologists and also scholars from the humanities, social sciences, political sciences, law and economics agreed in that 'the critical issue most commonly called "population control" is not *whether* the number of human beings shall be limited, but *how* the limitation should occur'.[50]

Time and again, 'the finite size of the earth, as graphically demonstrated in photos from the recent Apollo missions', was quoted. The limited earth assured 'that man's numbers cannot increase indefinitely'.[51] The new 'one world' view of interrelatedness entailed the expansion of programmes for a collective and uniform world development and involved population as a field for international technical aid. In 1969 the United Nations Fund for Population Activities (UNFPA) was established. The organization financed 'population assistance' particularly in the 'developing world'. 1974 was announced as the World Population Year. The UN conference on the Human Environment in Stockholm in 1972 fashioned the One Boat concept, the thought that all of humanity shared a common fate within absolute limits, derived from the motif of the ship harbouring the community of mankind.[52] Population, said Duden, was transformed from an exogenous into an endogenous factor of a 'world system' whose 'survival' seemed at stake.[53]

'Human ecology' became an integral part of the systems sciences that were emerging in the same time period. Human ecology explored human populations as an element of global ecological relations in the perspective of closed systems analysis. Human ecology was only partly indebted to the tradition of demography, population statistics and population politics that had developed in the national economies of the nineteenth century. Rather, as seen in the work of Pearl, human ecology was developed from biology and population ecology, which had traditionally focused on animal and plant populations. The fact that ecologists turned to human populations as scientific objects, combining evolutionary theories, numerical ecological concepts, demography and statistics demonstrates how strongly the natural sciences engaged in the discourse of population growth in the 1960s and 1970s. Human ecologists combined developmental and environmental dimensions of population growth to present a solution to the pressing global concerns of the time: 'Too Many People – Too Little Food – A Dying Planet', or short: 'population – resources – environment'.[54]

Quite literally, ecologists addressed the population–resources–environment conjunction as sharing a common denominator. In the early 1950s, Fairfield Osborn, one of the leading US conservationists of the mid-twentieth century, revoked the Malthusian 'law' when considering the world 'under the control of the eternal equation – the relationship between our resources and the numbers as well as the needs of our people'.[55] According to Osborn, this relation

> finds expression in a simple ratio wherein the numerator can be defined as 'resources of the earth' and the denominator as 'numbers of people'. The numerator is *relatively* fixed and only partially subject to control by man. The denominator is subject to substantial change and is largely, if not entirely, subject to control by man.[56]

Arranging 'numbers of people' and 'resources' in a basic ratio, a fraction, opened up new perspectives of managing the predicament. The population problem was framed as a veritable accounting problem, to be solved by mathematical competence and numerical methods. Osborn concluded: 'We have now arrived at a day when the books should be balanced. But can they be?'[57]

Both ecological concepts and numerical-economical methods needed to be combined in the task of balancing the books and establishing ecological equilibrium. Facing this tremendous task, the logistic curve provided the perfect means to project the earth as a closed system and enrol the population–resources relation into systems dynamics with the aim of achieving homeostasis. In the hands of human ecologists, the curve became a mathematical tool as well as a powerful visual device to assemble individual subjects to an abstract large-scale world population with an obvious common destiny. The logistic growth curve gained performative strength from its smoothness and symmetry. Its perfect form and high aesthetic quality suggested truth, universality and eternity. Uwe Pörksen calls such types of images 'visiotypes'.[58] The concept of visiotype stresses the typecast amalgam of text and image able to translate quantitative knowledge into (self-) orientation and regulation, and capable of setting standards of public perception of global reach. Moreover, visiotypes allow for simple equalizations. They facilitate the formulation of 'global equations' from which political and social contexts are excluded.

Visiotypes are social tools; they establish abstract general denominators to assemble large and sometimes incoherent parts of society, and they display future movements, or trends. They are constituted in a way that makes them represent the 'nature' of things.[59] The 'natural law of population growth' and the S-shaped curve perpetuated the logic they described; they created a world population facing the imminent doom of entropic regression and death, and they equipped the narrations and accounts of the time with a definite shape.[60] The curve in particular drew together a world community out of fear of the utmost earthly limits.[61] Pörksen suggests that visiotypes express temporal development; in such curves one can hear the time ticking.[62] Looking at the population growth curve, one could literally hear Ehrlich's 'population bomb' ticking away. His essential message, 'The population bomb keeps ticking',[63] made population growth seem like an imminent catastrophe, and possibly the work of a deliberate aggressive act. The drawing of a bomb on the book's front cover vividly supported the message. Its burning fuse conveyed that there would be no delay; the explosion was to be expected immediately.

From the empirical world population growth curve a variety of threatening trend curves were extrapolated. Most of these curves presented some variation of the logistic curve in the first half and projected quite disparate paths in the second half, giving rise to heated debates. From the beginning the logistic curve had been fraught with the problem that it represented only a single cycle of population growth within a definite time span, leaving open the question of how a population would develop following its phase of stagnation. In 1972 the study *The Limits to Growth* took up this question, putting new emphasis on world population as a whole. In the first chapter on 'The Nature of Exponential Growth', the authors took pains to reaffirm the first half of the logistic curve. From there they developed different future scenarios that they called the 'behavior modes of the population-capital system'. These behaviour modes translated the population–resources–environment equation to systems sciences, and according to the study they were visually represented by four basic types of 'survivorship curves' that depicted the principally possible changes of population over time. A population growing in a limited environment could approach its 'ultimate carrying capacity' – the asymptote of Pearl's logistic curve – in three ways: it could 'adjust smoothly to an equilibrium below the environmental limit by means of a gradual decrease in growth rate' (which would be Pearl's version), it could 'overshoot the limit and then die back again in either a smooth or an oscillatory way', or it could 'overshoot the limit and in the process decrease the ultimate carrying capacity by consuming some necessary nonrenewable resource'.[64]

Computer pioneer Jay Forrester provided the 'world models', the computer programs to model 'world dynamics' for *The Limits to Growth*.[65] Similar to Pearl, Forrester exhibited a cavalier attitude towards empirical information. Paul Edwards noted how Forrester put his models first, subsequently fitting them with highly aggregated data.[66] Like Pearl, Forrester presupposed 'life cycles' of his modelled entities or, in systems sciences' terms, 'behaviour modes' of growth, stagnation and equilibrium states. And like Pearl, Forrester presumed that the systems he modelled were closed systems. New in his computer-based models was the practice of building on concepts of a complex interrelatedness of the social and the natural, the living and the artificial, overriding traditional dualisms. These concepts in turn enabled Forrester to make strong predictions of a system's future and to open up visions of a system's management. Forrester, like Pearl, believed in the universality of his practices and his results. But his technoscientific infrastructure provided a more compelling foundation to the claim that different entities could be handled in the same terms. While early population ecology had reduced human social relations to their reproductive quality, systems ecology now in return upgraded these human breeding groups into computable socio-natural systems. Within these new systems, terms like 'reproduction' and 'behaviour' denoted both biological and non-biological modes of replication, of amplification and of control.

As discussed in the previous chapter, when modelling the behaviour mode of the entire world system for *The Limits to Growth*, the MIT group labelled their computer plot based on unrestrained world development the 'standard run'. Population was one of the five major variables determining the results of the standard run. The behaviour mode of this world model turned out to be 'that of overshoot and collapse'.[67] The term 'standard', however, no longer seemed to denote a possible or probable trend but a prediction: the curve's performance prescribed a scenario of rapid growth and decline. Following Jürgen Link's reflections on the normalizing power of graphs and curves assembled from aggregated data, the standard run represented a characteristic scenario of economic affluence; it carried the thrill as well as the terror of irreversible 'denormalization' that is expressed in all modern aggressive 'trend stories' of boom and bust.[68]

The economic implications of this visiotyped 'trend story' of probable excess and possible moderation need to be grasped to understand the power and authority of the study *The Limits to Growth*, with regard to its political impact as well as to the prospects of humanity mastering its precarious situation. The report proposed that expected low world *performance* could be turned back into the normal state, to the 'State of Global Equilibrium', through expert intervention.[69] In the following sections I will explore the power relations embedded in expert economic and ecological rationality by discussing measures for population control human ecologists proposed to manipulate the world's standard run.

ZPG: Population Accounting and Accountability

'The Challenge of Overpopulation', Karl Sax maintained, was that 'nearly two-thirds of the world's people live at little above subsistence levels; yet these are the people who have the highest birth rates'.[70] Population growth, while considered a global problem, was not conceived of as affecting all human beings in the same way. The 'developed' nations felt most threatened since they had most to lose. Accordingly, they considered themselves in a position to decree what measures were to be taken. Reiterating the Malthusian claims, Sax recommended either 'positive checks' – high death rates – or 'preventive checks' – low birth rates; one of these two restraints, he claimed, would be needed to control population growth and effect the 'Demographic Transition'. This model, developed around the 1930s and expanded after World War II, proclaimed a shift from high to low birth rates and death rates for highly industrialized – synonym to mature and highly civilized – countries. Total population numbers would follow the S-shaped logistic curve, asymptotically approaching a relatively high but stable population just below the carrying capacity of the human environment. Constant growth rates would eventually give way to constant birth rates.[71] Not only for the 'undeveloped countries' but also for the industrialized economies of the North immediate and aggressive measures were called to enforce demographic balance, in short: 'Zero Population Growth'.[72]

To return to Kingsland's observation, the population problem involved figures and rates, and the answers came from mathematics and economics. Kenneth Boulding, inventor of Spaceship Earth and early proponent of ecological economics, suggested employing market mechanisms to avoid the 'Population Trap' and lower the birth rate in the United States. In 1964 he recommended a 'system of marketable licenses to have children' which would allocate birthrights to potential parents.[73] The smallest unit in his economy of reproduction would be the 'decichild'. Certificates would be tradable; the 'accumulation of ten of these units by purchase, inheritance, or gift would permit a woman in maturity to have one legal child'. Boulding imagined that women would also be able to accumulate the necessary certificates through marriage. Each person, both male and female, would receive eleven decichild certificates at birth to secure a birth rate of 2.2, or whatever number would warrant the desired reproductive rate of 1.0, that is: zero population growth. The market would set the price of the certificate, reflecting the general desire in a society to have children.

Boulding's was a forthright economic suggestion built on capitalist market rationale. Ehrlich argued similarly in his book *The Population Bomb* in 1968, which is an impressive example of how the discourse on environment and population grew more radicalized among ecologists in the 1960s and 1970s. Ehrlich's gloomy prophecy for the world foresaw either 'population control or race to oblivion'.[74] The 'Population Bomb' as well as the term 'Population Explosion' he had borrowed during the student revolts of 1968 from a leaflet that had been circulated in the 1950s. In 1954, at the peak of the Korean War, the businessman Hugh Moore had published a brochure titled 'The Population Bomb' and had it distributed in the US in millions of copies.[75] In Ehrlich's opinion, 'all [problems of the world at present] can be traced easily to too many people'.[76] Purporting that population was exploding 'on a finite planet' he warned that the limits to growth would be felt quite abruptly by the end of the twentieth century, and most likely within the next decade. Consequently Ehrlich attacked the growth paradigm that dominated the Western world, according to which GNP, the Gross National Product, was the measure of all things – 'as gross a product as one could wish for'.[77]

Paul Ralph Ehrlich, professor at the biology department at Stanford University in California and a specialist in entomology and population biology helped turn ecology into an academic field that integrated humans as objects of biological study. Ehrlich aspired to analyse the population–environment relation as part of a complex ecosystem, using the methods of systems analysis and operations research. From the physics of closed systems he borrowed the principles of thermodynamics, particularly the first law, which held that energy could not be transformed into its various forms without loss to heat and statistical confusion: entropy. Unavoidably, the world was heading towards its dreaded entropic state – heat death. Economic growth and population growth within a finite world would lead to increasing resource use. 'Resource depletion', Ehrlich predicted, would be the certain result, a situation that could be solved neither by classical economical principles nor by market rationales. In his opinion, to speak of the 'affluent society' was thoroughly misguided; instead, one should speak of the 'effluent society'.[78] Ehrlich's vision was one of ecological equilibrium; it was the vision of 'getting the earth back in balance'.[79]

In the naturalist analogy to a cancer as 'an uncontrolled multiplication of cells', Ehrlich understood the 'population explosion' as 'an uncontrolled multiplication of people'.[80] Based on the principles of natural selection in evolutionary theory, he formulated an argument of humankind's 'urge to reproduce' as the eternal evolutionary-caused burden of mankind, recalling Malthus's postulate of the 'passion between the sexes' and explicitly referencing Pearl's 'laws of population growth'.[81] On the basis of the principle of differential reproduction, according to which some genetic type would simply be 'outreproducing' others, Ehrlich subsumed cultural historical changes under larger naturalist schemes of development.[82] Following a moderately deterministic Malthusian logic of agricultural and industrial development he explained the situation in which the death rates had fallen below the birth rates, so that the population had produced 'surplus'.[83] He distinguished only two options to cope with the population problem: the 'birth rate solution', a societal restriction of birth numbers that he urgently recommended, and the 'death rate solution'.[84]

'Too many people – that is why we are on the verge of the "death rate solution"'. If measures were not taken soon, this solution would, in the form of war, diseases and mass famines, inevitably meet humankind in the decades to come. The 'war on hunger', announced by US President Lyndon B. Johnson only a few years earlier, had already been lost.[85] In Ehrlich's versions of the death rate solution that he called the Apocalyptic Horsemen, worldwide political conflicts combined gloomily with environmental catastrophes. Fast-spreading viral diseases and global epidemics like a 'super flu' would annihilate entire populations until all social systems collapsed and the human species became extinct.[86] To pave the way for the birth rate solution, Ehrlich called for modernizing the unsuccessful governmental programmes of family planning; individualist forms of birth control should be replaced by overarching population politics. He requested the introduction of drastic measures of regulation, so that – after a controlled 'dieback'[87] – world population would stabilize on a level of two billion people. From the historian's perspective, Ehrlich's statements of what had to be done to reach 'a stable optimum population size' are among the most problematic parts of his work. Next to financial incentives and sanctions in tax legislation he proposed a temporary forced sterilization of the US population by adding sterilizing agents to drinking water and food.[88] Oddly, the proposed measures against 'reproductive irresponsibility' allied with some of the emerging liberal ideas of the 1960s: simplifying the right to adoption, guaranteeing the right to abortion and voluntary sterilization, modernizing sexual education in schools and challenging Catholicism to rethink its position on unborn life and contraception.[89]

Lifeboat Ethics: The Moral Economy of Population Ecology

'Coercion? Perhaps, but coercion in a good cause', Ehrlich granted in reviewing his own propositions.[90] Ehrlich was one of the three original incorporating members of the activist group 'ZPG – Zero Population Growth' founded in 1968 to raise public awareness of the population problem. The organization sought to confront the US white middle class with its lifestyle of using up far more than its global share of natural resources and adding more than its share to environmental pollution. The initial mission was to encourage citizens to reduce family size: 'Stop at Two', 'Stop Heir Pollution' and 'Control Your Local Stork' were some of ZPG's catchy slogans advertised on bumper stickers, flyers and posters, in public service announcements, magazines and organized protest marches. In the first year the popularity of Ehrlich's *Population Bomb* briefly boosted group membership to more than 30,000.[91] ZPG founded its own 'Population Education Department' and in 1975 formed a 'Population Education Program'. The department produced classroom texts and a video titled 'World Population' that was used as an educational tool in public exhibitions, museums and zoos. ZPG did not confine its actions to showing movies and handing out condoms. The organization also urged changes in population policy and abortion legislation, and it opened vasectomy clinics.[92]

In 1971, three years after the founding of ZPG, the movie *ZPG – Zero Population Growth* presented a drama about the overpopulated earth in the near future.[93] Having children has been ruled strictly illegal for the coming thirty years. The address of the president of the 'world society' who announces this global decree to the citizens in a TV broadcast sounds like a direct response to Ehrlich's suggestions:

> My fellow citizens: It is with a heavy heart that I bring you the findings of the council. After deliberating in continuous sessions for the last four months in unceasing efforts to find a solution to the devastating problem of overpopulation threatening to destroy what remains of our planet, the World Federation Council has considered and rejected all halfway measures advanced by the various regional scientific congresses. We have also rejected proposals for selective euthanasia and mass sterilization. Knowing the sacrifices that our decision will entail, the World Council has nevertheless reached a unanimous decision. I quote: 'Because it has been agreed by the nations of the world that the earth can no longer sustain a continuously increasing population, as of today, the first of January, we join with all other nations of the world in the following edict: childbearing is herewith forbidden.' To bear a child shall be the greatest of crime, punishable by death. Women now pregnant will report to local hospitals for registration. I earnestly request your cooperation in this effort to ensure the last hope for survival of the human race.

Director Michael Campus set this movie in an unidentifiable, thickly polluted metropolis in the near future. The place only dimly recalls Richard Fleischer's New York City in 2022. Whereas Fleischer pictured the noisy and cluttered squalor of a crowded megacity, Campus portrayed a highly artificial and technologized society of uniform lethargic people living sterile lives indoors. Oliver Reed and Geraldine Chaplin play the young couple Russ and Carol upset with the condition of having to order a surrogate robot baby. Obsessed with the desire for a child of their own flesh and blood, they revolt against the edict that has been operative for eight years. Carol secretly gives birth to a child that the couple hides carefully from friends and neighbours. Eventually, however, the young family is discovered by a neighbouring couple, itself with a strong desire for a child. While consenting at first to share the child, soon jealousy spoils the deal. A fight about proprietary rights ends in blackmail, betrayal, finally disclosure of the child to the authorities, arrest and approaching elimination. Still, the movie ends on a small note of hope. By way of the junk-littered canals beneath the city, the original family escapes in a tiny lifeboat. A rubber dinghy takes them to an abandoned beach. Ruins still discernible in this bleak landscape give a faint glimpse of the environmental devastations that were carried out in the name of world peace in the 1970s. The audience learns that until 1978 the site had served as a trial site of the nuclear 'Polaris' missile. In the former radioactive zone the family sets out to make a new start.

    In his work Ehrlich likewise employed the figure of the ship, though not as a symbol of survival and new beginning, but as a symbol of the threatened earthly community. By applying the image of 'the good ship Earth' he denoted the 'ever-shrinking planet' and its exceeded 'carrying capacity'.[94] His view of the world was that of a community of intricate interdependencies, thrown back to its own resources, that could not rely on any safe haven elsewhere. 'It is obvious that we cannot exist unaffected by the fate of our fellows on the other end of the good ship Earth. If their end of the ship sinks, we shall at the very least have to put up with the spectacle of their drowning and listen to their screams'.[95] In the same breath Ehrlich discussed international developmental aid as a distribution problem to be rationally managed, while acknowledging that developmental aid also served US hegemony in a world of decolonization: 'It might be to our advantage to have some UDCs [underdeveloped countries] more divided or even rearranged, especially along economic axes'.[96]

'I know this all sounds very callous, but remember the alternative'.[97] Ehrlich's neo-Malthusian assessment and solutions to the 'population problem' remain contested to the present day. Some reviewers criticized Ehrlich on his own terms, questioning numbers and statistical methods or contesting the geometric growth model, while others pointed out that reasons other than population growth should be held responsible for poverty, famines and economic regression.[98] Some opponents altogether denied the dismal consequences of a rapidly increasing population, claiming that progress in technology and agriculture would be capable of breaking even the costs of the rapid growth of human numbers.[99] But the charges of fascist traits in Ehrlich's work are as shortsighted as the arguments of the redemptive powers of modern science and technology have been vain. In order to avoid repeating the contemporary fallacy of choosing between either technological proficiency or oppressive sufficiency, we must analyse how normative assumptions were inscribed into the 'natural laws' of population development that justified rational schemes of population control.

The concept of 'carrying capacity' became the appropriate measure of the earth's maximum population. 'Carrying capacity' was defined as the maximum number of a species that can be supported indefinitely by a particular area or habitat, without reducing its ability to support the same number of organisms in the future.[100] Similarly, Liebig's law of the minimum was redefined in systems terms as the 'maximum sustainable yield of a natural biological system'.[101] Based on such quantifiable earthly limits, the concept of 'overpopulation' stipulated the assertion to be able to identify, to study and to handle resulting population 'surplus'. Finally, Ehrlich transformed the population–resources–environment relation into a mathematical equation which held that 'Environmental disruption = population x consumption per person x damage per unit of consumption'.[102] A scientific structure emerged in which the question to be posed was no longer 'How Many People Can the World Support?' Instead the question was: 'What is the *optimum number* of human beings that the earth can support?'[103] To take up Foucault's apprehensions, the question was explicitly couched as a problem of *storage*. The search for an optimum population to achieve ecological balance in view of the earth's critical carrying capacity framed the world population problem as an accounting problem of efficiently allocating human elements to a limited cargo space.

Garrett Hardin, biologist and professor of human ecology at the University of California at Santa Barbara, became well known for his view of the earth as a storage space with limited capacity and his proposition to 'close the commons' in breeding.[104] In a legendary *Science* article of 1968 titled 'The Tragedy of the Commons', Hardin attacked ideas and practices of designating 'commons' like the air and the sea and assign them to the responsibility of the international community. The contamination and scarcity of any commonly owned natural resource, he warned, would necessarily increase, as it would inevitably be over-used within a limited world. In his opinion the same held true for the hitherto common access to procreation. Hardin demanded a 'fundamental extension in morality', in the Old Testament formula *Thou shalt not*: 'Thou shalt not exceed the carrying capacity' became his quasi-biblical commandment of ecological correctness in the 1970s.[105] He took on a godlike authority when demanding that society should oppose the 'present policy of *laissez-faire* in reproduction'.[106]

To Boulding's suggestion of an economical system of child certificates, Hardin acidly objected that its general effect 'would be to allocate child-permits as we now allocate Cadillacs – to the richest'. Nonetheless, he conceded 'the scheme might be a useful interim measure in getting people used to the idea of parenthood as a licenseable privilege instead of a right'. Hardin also criticized overly simple reproductive schemes: 'I'm afraid there are more patterns of marriage and sex than are dreamt of in Doris Day's philosophy'.[107] Human ecologists most often built on a heteronormative matrix, acting on simple assumptions of two unequivocal sexes and corresponding categories of gender, gender identity and sexual orientation to conceptualize the growth of populations in terms of a male–female reproductive nucleus. Also, human ecologists did not take social and political relations serious on their own account or develop utopian forms of society, but based their image of social systems on established biological principles. In his 1972 book *The Voyage of the Spaceship Beagle* Hardin employed the name of Darwin's research vessel *Beagle* to paint a dystopian image of a human society on an expedition through outer space. The spaceship was his metaphor of an earth made vulnerable by the ethics of a population freely reproducing and exploiting their limited commons. At the same time the setting of a science fiction fantasy allowed him to explore various devices for reassigning economic responsibilities to regulate not only the use of the modern commons but also the size of population in a Darwinian fashion.[108]

Still, it is important to reflect on how the global schemes Hardin and other population ecologists proposed were different in motivation and in direction from the antinatalist and pronatalist policies that had been enforced in Europe and North America earlier in the century. Unlike social and racial anthropology and eugenics, the ecological approaches to the 'population problem' of the late 1960s and early 1970s focused not on particular groups or nations but on 'world population'. Solutions were derived not from theories and practices of heredity to ameliorate and optimize populations but from the goal to secure the 'survival' of the planet that stressed overall ecological and economic affluence. Nevertheless, it is no less important to realize how propositions involved similar principles of classification, designation and evaluation. The taxonomies resulting from aggregating and disaggregating human beings on the global scale identified certain regions and populations as less valuable and more threatening to the well-being of the human collective than others.

Hardin's demand to grant reproductive privileges to creditable individuals and nations only, strictly on the grounds of ecological reason, reveals the rational and the moral economy operative in human ecology. Lorraine Daston introduced the term 'moral economy' to science studies to discuss the set of values regulating a specific scientific culture.[109] I suggest employing the concept to unfold the implicit and explicit values and evaluations in the theories, claims and accounts of human ecologists. Based on the earth's limited carrying capacity, human ecologists proposed perfectly reasonable schemes to decide on who was to live and who was to die on a global scale. These regulatory suggestions were designed to overcome the 'normal population cycle', which, according to the biological law of population growth, would continually produce overpopulation. Ultimately the framework stabilized the biological construction of a normal population, of normal growth and of dispensable surplus.

'Will they starve gracefully, without rocking the boat?'[110] Clearly, in Ehrlich's doubtful question 'they' referred to the underdeveloped nations on the verge of drowning, while the global 'we' applied to the wealthy nations, and above all to the Americans, who perceived themselves as being in charge of the planetary ship. In 1967 the brothers Paul and William Paddock, one a retired Foreign Service officer, the other an agronomist and plant pathologist, predicted that a dreadful global famine would occur in 1975. In order to prevent the approaching Malthusian food–population collision, the United States would have to distribute food to poor nations in a reasonable way. To categorize between deserving and undeserving nations, the Paddocks suggested adopting the strategy of medical first aid by applying 'triage', the classification of three used in military medicine to assign priority of treatment to the wounded for the purpose of saving the maximum number of lives.[111] Given limited medical facilities, the selection principle of triage arranges for the tragic group of the 'can't be saved' to be left to perish while the 'walking wounded' are expected to survive without immediate help. Priority of attendance is given to those most likely to be saved by immediate medical care. The Paddocks suggested executing triage on the level of nations to allow a rational choice between those nations doomed to suffer the Malthusian catastrophe, those nations able to cope with overpopulation on their own, and those nations that should receive food aid with the chance to overcome their crisis.[112] Perceiving the world situation as an economic problem of the wealthy nations to allocate their limited resource stocks, triage appeared as an objective, sensible and practical procedure. At the same time triage confirmed the hegemonic claims of the West. The Paddocks were more outspoken than Ehrlich: 'The time of famines can be the catalyst for a period of American Greatness'.[113] Notably, those countries considering themselves authorized and able to apply triage on a global scale were also the countries in the position of aggregating and disaggregating populations statistically, thereby consolidating the global classifications they identified.

To balance world population for survival, Hardin brought forth what he called Lifeboat Ethics. Lifeboat ethics altered the metaphors of the earth of 'one boat' and 'sinking ship' to the 'lifeboat', stressing neither hope nor unity but limited capacity.[114] Hardin argued against the metaphor of Spaceship Earth, since it assumed a captain or helmsman in charge of the ship, while the lifeboat merely emphasized the necessity of applying strict principles of selection to its occupants, regardless of status. Securing survival in a lifeboat depended solely on the rational allocation of provisions and the disposal of dead weight. On this ethical basis Hardin made his 'Case Against Helping the Poor', arguing against the 'fundamental error of the ethics of sharing' in international-aid programmes and urging wealthy nations to close their doors to acts of charity like immigration and food aid to the poor. 'No one who knew in his bones that he was living on a true spaceship', he contended, 'would countenance political support' of the population of those countries who would simply 'convert extra food into extra babies'.[115]

In Hardin's view, the *optimum* world population, the population that would be able to survive on the planetary lifeboat, would have to be reached via a Darwinian process of selection that reflected a nation's 'fitness'.[116] Hardin defined fitness according to classical liberal logic of achieved economic prosperity. His disposition of human lives left the historical roots of disparities of wealth through colonial history and postcolonial power relations unnoticed and untouched. Socially and historically developed problems he defined as biological in origin and individual in character. The classification into top nations to be rewarded and nations not able to economize cleverly to be punished by suffering heavily suggests that Hardin like many of his contemporaries was not heading for a political solution of the problem of 'too many'. Nor did he aim at a fair international distribution of power and resources. The 'good cause' in whose name Ehrlich, Hardin and the Paddocks claimed to act, deflected from its inherent genocidal logic. When pointing to the growing disparities they aimed primarily at a balance of costs and benefits that would sustain the political and economic supremacy of the 'first world'.

An international food bank, Hardin maintained, would be simply 'a disguised one-way transfer device for moving wealth from rich countries to poor'. 'The demoralizing effect of charity on the recipient has long been known'.[117] Leaning on the basic survivorship curves proposed in *The Limits to Growth*, Hardin argued that without conscious population control a nation's population would endlessly repeat the cycle of overpopulation followed by a drop-back to 'the "normal" level – the "carrying capacity" of the environment – or even below'. If such countries were able to draw on world food-bank resources in times of emergency, he cautioned, the normal cycle would be replaced by the 'population *escalator*': The input from a world food bank would act as 'the pawl of a ratchet', pushing the population upward.[118] In a modification of the basic survivorship curve, Hardin's population escalator was composed of a continuation of S-curves that Jürgen Link has called the 'endlessly growing "snake"'; this curve carried the 'thrill' of an irreversible 'denormalization' through unsettling and thus immoral acts of charity.[119] The 'ratchet effect', Hardin warned, would end only with the total collapse of the entire system. 'The crash is not shown, and few can imagine it'.[120]

Carrying Capacity: The Good Cause of Ecological Intervention

Soylent Green is set in New York City in 2022, where most human beings alive are considered 'surplus'. Particularly old people are encouraged to consent to euthanasia. Edward G. Robinson plays the aged Jewish savant and registrar Solomon Roth who, like the astronaut Freeman Lowell in the movie *Silent Running*, establishes the connection to the past. Sol holds fond memories of a long-gone era when flowers, fresh fruit, vegetables and meat were a part of everyday life. The movie references both Nazi and ecological genocidal logic when Sol decides to end his life in one of the governmental euthanasia facilities. The beauty of the past resurges again in the classical music and in the images that accompany him in the moment of his parting. Surrounded by Edvard Grieg's *Peer Gynt Suite* and Ludwig van Beethoven's *Symphony No. 6*, Sol reviews a lost world in filmic projections of clear waterfalls and rolling waves, green meadows and thick forests, wild flowers and animals, magnificent mountains and spectacular sunsets.[121]

Stanley R. Greenberg wrote the screenplay for the movie, and he made some crucial amendments to the theme of Harrison's book. Charlton Heston plays the main character Robert Thorn, a detective on the New York City police force who investigates the murder of a board member of the Soylent Corporation. In the course of his enquiry, Thorn discovers a perfidious scheme of waste recycling. The company devised a most efficient way to handle the great numbers of human corpses. As living beings are considered excess, the dead are outright waste, their bodies disposed of by garbage trucks. At the same time this waste has come into focus as productive organic material that can be recycled and reintegrated into the food chain. Thorn watches Sol's body being delivered to a plant from where it is returned in the form of nutritious green pellets: Soylent has created a most efficient and profitable industry to transform the dead into food for those still alive. 'Soylent Green is people!' Thorn's desperate outcry at the end of the movie has acquired cult status among science fiction devotees.

To Ehrlich the day still seemed 'far away when food for billions is grown on synthetic nutrients in greenhouses free of pests and plant diseases, when the wastes of civilization are recycled entirely by technological means, and when all mankind lives in surroundings as sterile and as thoroughly managed as those of … an Apollo space capsule'.[122] Nevertheless, serious attempts were made in his time to chemically synthesize food, study artificial photosynthesis and mass-cultivate food substitutes like fungi or the single-celled algae named *Chlorella*, which made the leap to futurist visions like *Soylent Green* easy.[123] Although the closed industrial-organic metabolism was never fully implemented, the late twenti-eth century did devise new ways to control and regulate the lives and deaths of human beings on a global scale. Ehrlich eventually arrived at a practical equation to frame the population–resources–environment relation mathematically: The IPAT or 'impact formula' calculates the human *Impact* on the environment from the product of *Population* (number of people), *Affluence* (average per-capita con-sumption of resources) and *Technology* (inflicted environmental damage).[124]

My aim in this chapter was not to debate measures of population control as environmental strategy, let alone to find solutions to environmental problems. Nor could it have been my intention to resolve the ongoing debates between doomsayers and doomslayers. Here I have tried to draw attention to the ways in which the law of population growth, expressed by a simple numerical equation and visualized by a highly evocative graph, naturalized a particular biostatistical model of population development. Laws like Pearl's growth law and concepts like the earth's carrying capacity gave evidence for overpopulation and for the cat-egory of human surplus. Formulas like Ehrlich's impact formula helped to lock historically contingent situations into global equations that were held to be uni-versal and timeless. Such formulas acted as *prescriptions*, as instructions on how to perceive a certain historical situation. Based on effective calculus, they diverted from understanding how other approaches than those that can be scientifically expressed as equations and curves could be meaningful in assessing a problem.

In ecosystems science and literature, 'carrier functions' would become a cen-tral element of the earth's 'life support functions', while numbers of people were taken into account as the 'load'.[125] What had once been Malthus's principle of 'the passion of the sexes' was rephrased in ecosystems terminology: 'The cumulative biotic potential of the human species exceeds the carrying capacity of its habitat'.[126] Ecology and human ecology gathered social issues into natural laws and augmented the juridical authority of the (life) sciences, opening up new fields for biological intervention. In the case of world population, human ecology devised a new science of allocation. Hardin's claim of the evolutionary necessity to invest and disinvest providently in human resources with the aim to increase the wealth and well-being of the human collective corresponded with scientific models of ecological and of economical balance. Human life was summed up and totalled, evaluated and balanced. Both notions of balance served to monitor, circulate and allocate natural capital and human capital according to economic and environmental performance.

Scholars have cautioned against the specific ensemble of environmental concerns, actions, interventions and regulations of the late twentieth century, arguing that such 'environmentality' entails problematic dispositions of nature and (human) life in the name of an environment in need of protection.[127] The issue is not that ecology and population ecology were built on explicitly racist or eugenic categories. Ecologists like Ehrlich openly distanced themselves from such ideological biases of their professional field and its entanglements in the tradition of racial anthropology of the late nineteenth and early twentieth centuries. Instead it is critical to take note of the strata of scientific ecological concepts, which, phrased in neutral formulations, defined the course of things in a much more fundamental way than any political ideology. Of more concern than the scientists' divergences from a neutral and objective science is the notion that such a science exists at all.

5 CLASSIFICATION: BIOSPHERE RESERVES

One for One

Welcome to the twenty-third century. Three hundred years into the future, 'the survivors of war, overpopulation and pollution live in a great domed city, sealed away from the forgotten world outside. Here, in an ecologically balanced world, mankind lives only for pleasure, freed by the servo-mechanisms which provide everything.' By the year 2274 the major catastrophes of the twentieth century are history. Humans live in an entirely artificial environment under a closed biospheric dome. In their sparkling and amenable surroundings people are neither troubled by memory nor constrained by tradition or close relations. They are burdened neither by tedious duties nor by responsibilities. Instead, the inhabitants of this pleasant place are equipped with every gadget conceivable for leisurely distraction. Their existence is fun-filled and carefree, idle, sexually liberal and entirely self-indulgent.

The movie *Logan's Run*, released in 1976, presents a world in perfect eco-logical balance.[1] To maintain equilibrium life in this world is transient in every respect. Each citizen's visible life clock turns red at the age of thirty; he or she will die unless reborn in the ritual of 'Carrousel'. On their 'Last Day', people get the chance to try for 'renewal' in a public spectacle that recalls a Roman amphi-theatre. Carrousel imitates the authoritarian and paternalist logic of *panem et circenses*. The entertaining show serves to distract the audience chanting and ranting on the stands from noticing the rigid and merciless performance of pop-ulation control. Inadvertently the spectators witness the execution of an intricate scheme to which they themselves will be subjected without exception. The ser-vomechanisms which provide every amenity for a convenient life also control the beginning and the end of every human being. An elaborate cybernetic sys-tem controlled by a central computer constantly monitors and regulates every aspect of life under the dome.

To maintain ecological stability the domed city has been built on an invis-ible basement of channels and pipes, wires and cables, circuits and switches that ensure power supply, water supply, sewage, the cleaning and circulation of air and a working communication system. The central automatic feedback system that controls the interior environment also stabilizes the human population. Human offspring is created by a technology of 'seeding' *in vitro*, farmed in 'breeders' and raised collectively in age cohorts. The enclosed world keeps the total number of human lives stable according to a strict 'one for one' rule. In the course of their fleeting lives the citizens have come to take the principle of replacement for granted: 'Well, why not? That's exactly how everything works. Keeps everything in balance. One is terminated, one is born. Simple, logical, perfect. You have a better system?'

The Science and the Fiction of Replacement

From a present-day perspective, the ecological schemes drafted in the 1970s for deciding who would remain on the planetary ship named earth and who would count as surplus and would have to disembark may seem at best irrelevant and at worst aberrant. However, when considered as part of a theory of a maximum carrying capacity of a spaceship kept in balance, they are plausible and rational, and this is what makes them so disturbing. Lifeboat and spaceship economies not only created new metaphors for understanding the vulnerability of planet earth, they also favoured scientific and technological laws and principles over social and cultural values and beliefs. They instigated rigid classification schemes and a new ecological ethics for selecting life and nature on a global scale, with the aim of establishing a rational scientific basis for determining what would be useful and what was redundant, what was to be conserved and what was to be discarded.

Peter Sloterdijk points to the selectivity that characterizes all ark narratives. In all stories of the ark, he reminds us, the choice of the few is declared a holy necessity, and salvation is found only by those who have acquired one of the few boarding passes to the exclusive vehicle.[2] In the previous chapter I discussed some of the selection principles proposed in order to decide who would be accepted aboard the earthly spaceship. This chapter will explore how ecological principles of selection were put into practice. I will study the composition of the nature that was enclosed in Biosphere 2, a project launched in the Arizona desert in the 1980s to recreate the earth's biosphere *en miniature* in a great domed habitat. Sealed away from the outside world, the servomechanisms of Biosphere 2 maintained ecological balance. Taking up Sloterdijk's supposition of selectivity in enclosed habitats, I argue that the project of Biosphere 2 emphasized not completeness but systems integrity. The single parts to be taken aboard were selected according to the principles of ecosystems science: functionality, efficiency and replaceability.

Biosphere 2 embodied and performed the three principles of 'circulation', 'storage' and 'classification' that Foucault had anticipated in the late 1960s and that have provided the structure of this book. Circulation refers to the bio-spheric metabolism, to the circulatory system installed in Biosphere 2. Life in Biosphere 2 was grafted onto a complex technological foundation that was barely noticeable in the lush and fertile glass dome. Underneath the paradisiacal garden worked a machinery in order to provide life support. Classification points to the specific configurations of the organic and the technological. Classification concerns the ways in which the elements of Biosphere 2 were selected, and how they were meant to cooperate to form a whole. Finally, storage applies to the limits of the biospheric container. Storage informed the answers to the question of how the confines of the glass dome restricted the amount of life and technology to be fitted inside.

Biosphere 2 can be seen as the paradigmatic twentieth-century example of what Sloterdijk calls the 'ontology of enclosed space'. Sloterdijk's image of the 'swimming endosphere' was realized in the project of an indoor nature that was to provide the only possible environment for its inhabitants in the face of ecological catastrophe.[3] I will discuss Biosphere 2 as a storage space in which the inventory of the earth – in the form not only of species, but also of environmental knowledge and ecotechnologies – was assembled. Biosphere 2 was operated with the aim of reconstructing and eventually replacing the earth. I am interested in the constitution, the structure and the statute of the nature rebuilt under ecological emergency rule.[4]

I also suggest that the list of questions that Spaceship Earth raised in the Environmental Age needs to be amended. In previous chapters I have reflected on the contemporary debates of how many people the world could support and discussed how the problem of *how many* was framed in ways that expected a numerical answer. The notion of a maximum number of lives gave rise to the search for the optimum number of lives on the planet. This inquiry entailed the question of *who* would ultimately belong to the group of the most select. To *how many?* and *who?* I now propose to add *where?* Where was Spaceship Earth heading? I propose that Spaceship Earth was more than the image of a natural environment that had to be balanced like a fragile ship, or operated by human engineering skills like an intricate machine. Spaceship Earth has to be understood as an image of the earth as a self-sustained vehicle that could be moved and maneuvered in its entirety. The project of Biosphere 2 embodied the idea of Spaceship Earth as a modern ark. This modern ark was designed not only as a space of storage to wait out the ecological catastrophe, but also as a frontier vehicle that would transport the earthly habitat to formerly uninhabitable regions in the universe.

While Garrett Hardin's lifeboats continued to be fraught with terrestrial problems, Spaceship Earth projected the planet as a temporary environment, opening up the prospect of leaving the planet altogether. 'Men in a spaceship are not locked in one place, but become perpetual travelers', William Kuhns observed in 1971.[5] In 1980 Kenneth Boulding, in an afterthought to his original 1966 essay on Spaceship Earth, openly admitted to the colonization of space as a vision that was closely connected to the metaphor of the spaceship.[6] And Sloterdijk reminds us that the ark mythology includes the radical idea of completely removing the 'endosphere' from nature.[7] Biosphere 2 represented just such an artificial construct of a nature materially isolated from the outside world. The project aimed to facilitate the study of the earth's ecological systems on a small scale in order to permit the development of a self-contained and ultimately self-sustaining 'living' system to colonize other planets.

'Why not build a spaceship like the one we've been traveling on – along with all its inhabitants?'[8] The architect of the Biosphere 2 project, Philip Hawes, posed this question, demonstrating how the mythical narrative of the spaceship named earth had taken hold. As discussed in chapter 2, Roland Barthes argues that 'Myth hides nothing and flaunts nothing: it distorts; myth is neither a lie nor a confession: it is an inflexion'.[9] Indeed, Spaceship Earth was not simply a metaphor that concealed a reality or truth. Neither did Spaceship Earth perpetuate any fabricated falsehoods. Barthes has proposed that myth, when placed before the alternative, finds a third way out, a third way he identifies as 'naturalization'. Biosphere 2 naturalized the myth of Spaceship Earth; by essentializing the account of survival of threatened humanity Spaceship Earth entered the order of the natural. The spaceship materialized as the space under the glass dome.[10]

In the following sections I will discuss the concept, the goals and the design of Biosphere 2 and explore how it was put to work. I also explore the links between Biosphere 2 and US space politics of the 1970s and 1980s. I am particularly interested in NASA's plans to establish a permanent human presence in space subsequent to the Apollo Program. I will then study the economy of replacement implemented by Biosphere 2 with regard to the shifts that systemic substitution entailed for the meaning of 'life', 'life support' and 'survival'. The final sections explore how Biosphere 2 put into practice a notion of nature as a sanctuary and reserve. I discuss the environment of Biosphere 2 as a heterotopia, a site that was simultaneously an ark, a spaceship, a frontier community and a colony, designed to survive anywhere in the universe. Biosphere 2 was an exceptional nature with exclusive rights of access attached to it that determined who in a state of emergency would be able to use this nature and to control it.

Biosphere 2: Living Laboratory

In 1983, when Biosphere 2 was in its planning stages, the American economist Jeremy Rifkin thought two futures likely. Each presented alternative options in facing the challenges of the late twentieth century: the ecological future that would be well-balanced and well below the earth's carrying capacity, and the engineering future that sought to recreate the earth in humankind's own image, freed from the restraints of nature:

> Two futures beckon us. We can choose to engineer the life of the planet, creating a second nature in our image, or we can choose to participate with the rest of the living kingdom. Two futures, two choices. An engineering approach to the age of biology or an ecological approach.[11]

Biosphere 2 demonstrates how these future choices were realized under one single roof. In the Environmental Age the ecological future and the engineering future no longer presented mutually exclusive options. The second biosphere combined the concerns of a threatened earthly environment with the promises of modern and efficient technological environments – enclosed autarkic habitats, secluded, sanitary, versatile and fully conditioned.

The private enterprise started in the early 1980s on a site of three acres in the Sonoran Desert, about 25 miles north of Tucson, Arizona (see Figure 5.1).[12] In the years between 1984 and 1993, during its major phase of operation that this chapter will focus on, Space Biospheres Ventures (SBV) controlled Biosphere 2. The joint venture consisted of a private group of researchers, developers and investors assembled under the name of Decisions Team, contributing managerial as well as scientific and technical expertise, and the Decisions Investment Corporation, a business that supplied the necessary capital. The costs of building Biosphere 2 were estimated to be roughly thirty million US dollars. Most of the funding came from the Texan oil millionaire Edward Perry Bass, who owned about 90% of Decisions Investment. Moreover, two minor private sponsors supplied funding. One was the London-based company Synergetic Architecture and Biotechnic Designs (SARBID). The project was also supported by the Institute of Ecotechnics, formed in 1973, of which the leading figures in the Biosphere 2 project, John Allen and Mark Nelson, were founding director and chairman.[13]

Biosphere 2 consisted mainly of an enormous steel-framed enclosure inspired by the geodesic structures of architect Richard Buckminster Fuller, one of the designers of Spaceship Earth. Already in the 1960s Fuller had popularized the innovative spherical modular design that he called 'Noah's Ark #2'.[14] For the Expo pavilion of the United States at the Montreal World's Fair in 1967 he had constructed a impressive globular lattice of more than 60 meters in height that was made of individual triangular steel elements. Two years earlier, in 1965, Fuller had confronted the citizens of New York with the idea of doming-over a large part of the island of Manhattan for environmental and energy reasons. For this project he foresaw a comparable construction of steel frames and transparent glass or acrylic filling.[15]

Figure 5.1: Biosphere 2. Biospherians Before Experiment, 1990.
Philippe Plailly/Science Photo Library.

It was one of Fuller's collaborators, Peter Pearce, who designed the modular structure of the 'spaceframes' to carry the arching roof of Biosphere 2. In order to be able to survive in space, the structure was projected to last at least a century. To meet this demand of longevity, new materials, techniques and know-how were required. The problem involved the question of the right kinds of steel that would not only fit but that would endure the next hundred years, that would not leak out trace elements and that could carry a structure of up to 30 meters in height and a glass and acrylic surface of more than 15,000 square meters.

The specially designed panes to cover the steel spaceframes had to be transparent to permit sunlight to enter. At the same time they had to be airtight and watertight according to a one percent leakage standard to ensure that the new biosphere would be completely materially isolated from the outside world. A fully 'sealed' enclosure, however, posed the problem of balancing the fluctuations in air pressure within the dome. The erection of two large flexible expansion chambers or 'lungs' outside of the rigid construction solved this problem. One of the lungs served also as a reservoir of two hundred thousand gallons of water stored to compensate for the decreasing amount of free water on account of plant growth. Another difficulty concerned the control of the quickly rising temperatures under the sealed dome. The tropical greenhouse climate caused by shutting off air forced the designers to implement energy-consuming indoor ventilation and cooling systems. The energy intensity in turn proved so high that the plan of a purely solar-powered facility was abandoned. A power plant with three generators powered by natural gas and diesel fuel and a capacity of ten megawatts stood ready near the main facility.[16]

Under its enclosure Biosphere 2 housed an ecological 'mesocosm'. The large model ecosystem allowed unprecedented experimental ecological work at the scale of 'biomes' or ecological entities defined as 'large-scale complexes of life communities, soils and climates in characteristic geographic positions'.[17]Biomes represented distinct ecosystems that contributed the essential natural parts and processes to the 'major cycles of the biosphere' discussed in Chapter 3.[18] These miniature ecosystems were designed as 'living laboratories'[19] to study the behaviour of closed ecological systems and to promote the development of a self-contained and self-sustaining living system. Seven such individual biomes were laid out underneath the glass dome. The five 'wilderness biomes' encompassed a tropical rain forest, an ocean including a coral reef, a savannah, a mangrove-marsh biome and a desert. Next to the wilderness biomes an 'intensive agriculture' biome and a 'micropolis', a small 'city' for eight human biospherians were created. The wilderness and agriculture biomes were to cooperate in the effort to sustain the human habitat at the top of the food chain.[20]

The mesocosm of Biosphere 2 set up a new scale of ecological experiment to study the earth's ecosystems functions and interactions and to research the impact of humans and their technologies on the overall biospheric system, right along the line of environmentalist requests for a 'biosphere consciousness'.[21] Materially isolating Biosphere 2 constituted a new type of ecological experiment that incorporated human beings as research subjects and as research objects into a defined laboratory space. The laboratory was designed to spread sooner rather than later to encompass the entire earth.[22] The first closure experiments were carried out in 1987 in a sealed 'test module' of about 480 cubic meters that was furnished with small quarters to 'give life-support to one human being'.[23] These tests were 'unmanned', yet the experiment of Biosphere 2 was first and foremost a 'Human Experiment', to quote the title of Allen's book on the project published in 1991. Allen's viewpoint corroborates the suggestion that Rifkin's two separate futures had coincided: 'The paradigm of an ecological lifestyle had become so dichotomous between the technosphere and biosphere that many assumed that one had to choose between a future that was purely technological or one that rejected technics altogether'.[24] Allen proposed that a harmonious integration of both futures was needed to create a natural-technological living space. Within this hybrid habitat, the human econiche would be 'to function as intelligent managers and researchers of the biosphere, not as cave dwellers or science fiction characters in a lifeless world'.[25]

In September 1988 the first human closure experiment was carried out. Allen volunteered to be Vertebrate X, the first human to be locked into the test module for three days. Abigail Alling followed Allen in a five-day trial called Vertebrate Y in March 1989, and Linda Leigh became Vertebrate Z during twenty-one days in November 1989.[26] Two years later, the entire facility of Biosphere 2 was ready for the first scheduled closure period. On 26 September 1991, eight biospherians – four women and four men – were sealed into Biosphere 2 for the two-year closure experiment called Mission One.[27]

Life Support System

> With the dawn of the space-age came the view of Earth from space that enabled many people to see immediately, instinctively, that there was only one Earth and that it was finite and precious. The prospect of travel to other worlds began to make us realize that the biosphere of Earth was unique in our solar system and we would go nowhere off this planet for long *without a similar life support system.*[28]

The project managers of Biosphere 2 believed that the most powerful way to understand the earth's biosphere was to rebuild and to operate it. They framed their notion of the earth's life support system with the image of the spaceship, precisely by equating the earth's biosphere with the standards of technological life support utilized in space travel. To them it seemed principally possible that life support systems could be enhanced in such a way as to replace their original. Biosphere 2 was a project to understand, to maintain and to ship what had become life support on earth. Margret Augustine, CEO of SBV and co-designer of Biosphere 2, ascertained: 'As long-term self-regulating and sustaining support apparati, such biospheric systems will open the way for permanent habitation in space colonies'.[29] Biosphere 2 was designed as a transportable environment to secure human habitation across time and space, operable at any terrestrial or extraterrestrial location. Anticipating that in the long run life on earth would have to expand to other planets, the experimenters hoped to develop 'a prototype for a space colony' and to settle future 'sustainable communities' on Mars.[30]

In a small but visionary volume titled *Space Biospheres* and published in 1986, Allen and Nelson asserted that 'Biospheres are required for the US space programme to continue, even if only so far as the Moon and Mars, as well as for the long-term feasibility of permanently-manned Earth-orbit space stations'.[31] During the 1980s the managers of Biosphere 2 closely observed the proceedings of the US space programme. They strongly supported NASA's plans to build a permanently manned earth-orbit space station as a first step on humankind's way to Mars. And they welcomed President Ronald Reagan's endorsement in 1984 of the earth-orbital station *Freedom* that was projected for launch by 1992. The team displayed confidence: 'Space Biospheres Ventures drove to get Biosphere 2 built and into operation by that date [1992], anticipating the possibility of putting the first small space life system into orbit by 1995'.[32]

But the team of Biosphere 2 also rebuked NASA for its vague timeframe, complaining that political aspirations to extend earthly life towards the stars had weakened. Allen and Nelson pleaded for overcoming the 'present lack of overwhelming support for the space program'. In their eyes the US government had lost sight of the rationale for space exploration:

> But clearly there is something wrong. No second president [after John F. Kennedy], and all presidents are ambitious, has realized that he would make his place, the American place, and the human place in history forever by saying, 'We will make a settlement on Mars by such and such a date as the first step in extending Earthlife among the stars.'[33]

This was a reminder that planet Mars had originally been a goal of the American space programme. Indeed, while the endeavour of landing a human being on the moon in the 1960s continued, NASA harboured plans of constructing a reusable launch vehicle or 'space shuttle' that would eventually serve and supply a permanent earth-orbiting space station. Both space shuttle and space station were major components in the long-term goal of going to Mars.[34] In September 1969 the Space Task Group presented President Richard Nixon with a report on the different options for the US space programme in the post-Apollo period that included the alternative of a manned Mars mission for launch in 1986.[35]

Nixon had to take into consideration that public support for manned space flight had waned once the goal to land a human on the moon had been reached. The post-Apollo period of the 1970s in the United States was marked by matter-of-factness rather than by lofty visions. The Apollo 17 flight that ended the Apollo program in 1972 initiated what has been called the 'age of pragmatism'[36] in space flight. The early 1970s were a time of disenchantment in many regards. 1972 was a crucial year regarding the articulation of environmental concerns; 1973 saw the oil price shock in the Western world as well as a massive budget problem and a sobering end to the Vietnam War in the United States; and in 1974 the Watergate scandal forced Nixon to resign ingloriously from office. Amid the climate of general gloominess and public distrust against governmental agencies, the space programme continued to shift its work to less spectacular and less expensive projects of unmanned space exploration. In 1975, unmanned space probes were launched on their flight to Mars.[37]

Mars thus remained the distant long-term goal of space travel, but the idea of a permanent human presence in space was fostered by more practicable short-term resolutions. 'I have decided today that the United States should proceed at once with the development of an entirely new type of space transportation system designed to help transform the space frontier of the 1970's into familiar territory, easily accessible for human endeavor in the 1980's and 90's'.[38] In this announcement made in January 1972, Nixon reiterated Kennedy's image of space as the new frontier and Frederick Jackson Turner's image of the frontier as a recursive process. The first frontier posts were built not on the Moon or on remote Mars but in near-earth orbit. Nixon's decision directed the construction of the space shuttle as part of the new 'low-cost, multi-purpose space missions' the United States were settling for.[39] NASA administrator James C. Fletcher commented: 'The Shuttle is the only meaningful new manned space program which can be accomplished on a modest budget'.[40]

Skylab, the first US space station, was launched in 1973. Its construction and operation adhered to the low-budget plans by using up the available equipment left from the Apollo program. The 'sell-off' of spacecraft and rockets and the wholesale recycling of the Apollo hardware proved to be a profitable move.[41] At the same time Skylab represented a move towards space science. The exploration of distant objects by unmanned space probes as well as the distant goal of establishing a permanent presence in space led to space flight being perceived as a predominantly scientific venture. Compared to the earlier valiant missions of Gemini, Mercury and Apollo, spacefaring lost some of its former heroism and much of its extravagant playfulness. As an object of science, space flight was seen as sober and sound work. The space shuttle and orbital space station projects took up the fundamental question if and how permanent human life in space was possible. Skylab was about 'habitability'. The Skylab 1 mission featured a 'habitability support system'.[42]

The experimentalization of habitability in space required not fighter pilots but 'scientists-astronauts'. Skylab was a space laboratory to be visited by different crews of trained scientists for time periods of one to three months.[43] During three separate visits over 1973 and 1974 a team of three worked and 'lived' in the station. The two-story 'house' was considered reasonably comfortable and spacious. In microgravity the station was provided with a pressurized earthlike atmosphere in which spacesuits were not needed. Amenities included a galley kitchen and wardroom with a table for group meals, private sleeping quarters, waste management facilities, a shower that was custom-designed for use in weightlessness, a small library and a large window. The public imagination of the interior as a cosy enclosure completely submerged in a hostile environment compares to the fiction of the comfortable interior of submarine *Nautilus* in Jules Verne's story of ocean conquest.

The 1980s saw the realization of the long-planned space shuttle programme with the start in 1981 of the shuttle *Columbia*, which compared to the Skylab space station was perceived as a luxury hotel. The shuttles were equipped with a galley kitchen, lavatories, sleeping quarters and laboratories.[44] When Biosphere 2 was in its planning stages, the opinion was widespread that the United States and the Soviet Union had proven that it is possible to create artificial living spaces in outer space. In this race the Soviet Union had again come first. In 1971, two years ahead of Skylab, Russia had placed the space station Salyut 1 into orbit. In 1986 Salyut was replaced by the space station Mir.[45] Allen and Nelson met such demonstrated Soviet strength with awe and sarcasm: 'Little wonder the Soviet spacemotto is "We must grow our own apples on Mars!"'[46] Their reminder of the Soviets reporting good experiences with human physiology in microgravity environments was a reproach directed against the US government. 'In America, certainly all that is lacking is the political will to say, "We are going to Mars."'[47]

Prototype for a Space Colony

The 1980s witnessed a new vigour and urgency in US space exploration that overruled the pragmatism of the 1970s. In his State of the Union message in January 1984, Reagan pledged Congress and the American public to the project of space station Freedom to master the 'next frontier: space'. 'Tonight, I am directing NASA to develop a permanently manned space station and to do it within a decade'.[48] Obvious is his reference to Kennedy's Special Message to the Congress of 1961 that committed the nation to landing a man on the moon within a decade. In October 1984 Reagan established the National Commission on Space by executive order, issuing a report titled 'Pioneering the Space Frontier' in 1986. The report formulated a visionary agenda for the civilian space enterprise in the twenty-first century and announced the long-term exponential growth into future permanent space settlements as an overarching goal. Next to new propulsion systems and the establishment of 'initial outposts' to access the solar system, the commission recommended the development of 'fully self-sustaining biospheres independent of Earth'.[49]

Nevertheless, the Biosphere 2 managers criticized the existing and envisioned NASA space vehicles for their lifelessness and artificiality. They challenged NASA's space programme to consider such encompassing forms of life in space as offered by the 'space life system' of Biosphere 2:

> SBV's vision of a *created biosphere* was 180 degrees from what they were thinking about at NASA. The space agency's idea of a space station was of high-tech 'cans' containing lots of plastic, metal, computers, and walls painted in psychologically approved colors. There was little room for the odors, textures, sights, and sounds of a complex living assemblage of plants, animals, and bacteria.[50]

This criticism overlooks that the agency had long been shuffling design studies of self-contained and self-sustaining environments that would enable a permanent presence in space. In 1975 a NASA symposium had discussed the possibilities of creating self-sustaining habitats in space that would draw energy only from the sun – materially closed and fully recycled spaceships like the one Boulding had envisioned.[51] Controlled or Closed Ecological Life Support Systems, CELSS in short, denoted utopian design studies of system configurations that would provide not a mere minimum of life support functions but an environment able to produce food, water and air onsite, and to recover and recycle oxygen and water. These projects also pondered the kinds of plants and algae to be taken on board to decompose and regenerate waste.

NASA conducted studies on closed ecological life support systems from the late 1970s onwards. The agency had addressed the challenge of a fully autarkic miniature version of earth's ecosystem for future space travel as early as the 1960s: 'The world itself in all of its essential life-giving functions must be reproduced in miniature. The space craft and its passengers must eventually represent a completely self-sufficient community capable of an indefinite, independent existence'.[52] NASA's long-term plans were not far from Allen's biospheric vision; both foresaw the transportation of the earthly environment to space: 'Life might be sustained in space as it is on earth if enough of the "earth" becomes transportable. The task of miniaturization is to make more things fit into less space'.[53] In such imaginings of miniature transportable regenerative life support systems for future space colonization, the ark featured prominently as the figure of absolute self-containment and self-maintenance. 'Interstellar Arks' denoted the new dimension of space travel that accounted for more than one generation of duration. According to the truly visionary space travellers, it would not suffice to equip a spacecraft on an interstellar voyage with the most basic supplies for the transportation of a handful of discoverers. The space ark would have to serve as 'a mobile cradle of civilization – a self-contained small world, prepared for birth and death and for the cohabitation of many generations of humans that never saw their planet of origin'.[54]

In 1987 NASA's so-called Ride Report on 'Leadership and America's Future in Space' suggested establishing 'outposts on the Moon' that would 'represent a significant extraterrestrial step toward learning to live and work in the hostile environments of other worlds.' Moreover, the report proposed the bold program of 'Humans to Mars' that would advance human presence to an 'outpost on Mars' and include establishing a life sciences research program to validate the feasibility of long-duration spaceflight.[55] In 1988, the year in which the space shuttle *Discovery* set the United States back on track for space exploration after the two-year pause caused by the deadly *Challenger* accident, the crew of Biosphere 2 began the first closure experiments in the test module involving human beings. In the same year the NASA report titled 'Beyond Earth's Boundaries: Human Exploration of the Solar System in the 21st Century' envisioned 'the 21st Century as a time when humanity will have broken free of the physical and psychological bonds of planet Earth to live and work for extended periods on nearby bodies in our solar system'.[56]

On the occasion of the celebrations of the twentieth anniversary of the landing of Apollo 11 on the moon in 1989, President George H. W. Bush once more promised a program of manned exploration of the solar system and of permanent settlements in space.[57] In 1990, when Biosphere 2 was on the verge of being closed for its first two-year experiment, Allen had every reason to be optimistic: 'President Bush's third announcement during last year's celebrations, along with his plans to have Space Station Freedom in orbit by 2000 and a permanent lunar base soon thereafter, was his administration's commitment to a manned expedition to Mars by 2020'.[58]

Life-Enhancing Feedback System

For this moment Biosphere 2 was to be ready: 'SBV was on the verge of launching its own spaceship, so to speak, with eight people in it. It needed at least a rough draft of an operating manual'.[59] In plain reference to Buckminster Fuller's vision expressed in his 1969 book *Operating Manual for Spaceship Earth*, the assignment waiting for the project team concerned engineering and running a spaceship according to the superior design of life on earth. In the words of Allen and Nelson, the major motivation of the Biosphere 2 project was to 'create living art forms appropriate to the Space Age which celebrate the epic of evolution and which will produce heroes of a new kind – heroes who are champions of life and explorers of space'.[60]

> Although designed as a prototype for a long-term Mars colony, the feeling [inside Biosphere2] is not at all the cramped, clinical enclosure portrayed in science fiction. It has the feeling of a comfortable, livable community, achieving what astronaut Rusty Schweickart, at *The Human Quest in Space Symposium* in 1986, probably had in mind when he declared that the true challenge in space 'is not simply existing and surviving in space, but *living* in space'.[61]

Existing space stations as well as most of the drafted ones were self-contained but not self-maintained. They had to be serviced and supplied like an aircraft; they needed stocking, disposing and a permanent rotation of the crews, using spacecraft like the American space shuttles or the Russian Soyuz. The Biosphere 2 project aimed to build a robust ecological system capable of longer duration and supporting greater diversity than conventional life support systems in space. The idea was to build a settlement in space that resembled life on earth: 'Why not look at life in space as *a life* instead of merely travel?'[62] This proposal by the architect Hawes in 1982 opposed the dependency of astronauts in a spacecraft that had to be supplied from an earth base, a provisional state circumscribed by the image of the umbilical cord. The central issues were autarky and totality. Reciting the familiar images of environmental closure, from Sueß's Biosphere to Fuller's Spaceship Earth, Allen brings to mind the concept of wholeness:

The idea of the biosphere, thinking of the system as a whole, began to be discussed in more and more quarters. The name and emphasis varied – the biosphere, Gaia, Spaceship Earth, the global commons, the global environment – but the fundamental idea was the same.[63]

Notably, each of Allen's allegedly synonymous figures emphasizes different qualities of planet earth: the sphere containing all life; the organism with regenerative powers of spiritual quality; the cybernetic vehicle with a life support system run with technoscientific skills; the earthly resources as political entities common to all and proper to none; and the limited human living space on earth. Allen's symphony of terms tends to elide the observation that the different figures were discursively constituted differently and entailed different political choices. The fact that Allen could still claim that they shared the same 'fundamental idea' rests on the changes of the concept of 'life' in the process of describing the earth as a 'system as a whole'.

Both NASA's designs of future space stations and the design of Biosphere 2 were based on novel definitions of life at the turn of the 1970s. Derived from Austrian physicist Erwin Schrödinger's celebrated question about life posed in the early 1940s, new explanations of life emerged that relied on the contingencies of physical and biochemical principles.[64] The French geneticist Jacques Monod in 1970 gave a biochemical account of life as an autonomous system constructed not entirely coincidentally but intentionally as well, whose actions were purposeful and directed but nevertheless wholly contingent in existence, so as to create an organization capable of self-identical reproduction.[65] In an essay from 1967 the British physicist John Desmond Bernal conceptualized life as an open system of organic reactions favoured by catalysts, an explanation that based processes of life on thermodynamic principles.[66] In the 1970s, the new field of biophysics defined life as a highly organized system of material entities slowing down the ever-increasing entropy by means of autoactive and directed organization.[67]

By then life had lost any traces of *vis vitalis* prevalent in the late eighteenth and early nineteenth centuries. The creative life forces of animism and vitalism had been abandoned in the nineteenth century and replaced by the laws and principles on which biology and chemistry were centred. In the twentieth century, life's characteristics concentrated on structural order and functional complexity, on metabolic processes and endless reproduction according to the principles of biophysics, biochemistry and eventually molecular biology. In reference to Schrödinger's and Monod's apotheosis of chance in the evolution of life, in 1987 the German biochemist Manfred Eigen suggested to conceive of the stepladder of life as an evolving and increasingly complex regularity in matter that followed physical and chemical structuring principles to attain a state of dynamic organization and self-organization. His *Steps towards Life* were also meant to denote the scientific steps to be taken towards understanding life and evolution, from deciphering the blueprints of living beings stored in the functional entities of the DNA molecule since the 1950s to the optimization of their functional efficiency.[68]

BIOS – life – had also been the name of Russian experiments on closed cycles of converting carbon dioxide to oxygen and back.[69] In the early 1960s the structures basically sustained one human being and a small number of plant organisms: at the Institute of Biomedical Problems in Moscow Yevgeny Shepelev self-experimented with *Chlorella* algae. He spent twenty-four hours in a tightly sealed steel chamber called Bios-1. *Chlorella* regenerated his air and purified his water. The algae-based systems were further developed in the BIOS experiments in the later 1960s and 1970s at the Institute of Biophysics in Krasnoyarsk, Siberia. While test facilities Bios-1 and Bios-2 provided eight square meters of *Chlorella* per human being the Bios-3 facility reached the size of a small space station. Closure times of up to six months were attained with two and three-person crews, producing a balance of proteins, carbohydrates, vitamins and minerals. Food crops supplied up to 80 percent of the diet and provided part of air and water regeneration.[70]

An associate in NASA's science program on closed ecological systems was microbiologist Clair E. Folsome, director of the laboratory of 'exobiology' at the University of Hawaii. In the 1960s Folsome conserved miniature 'ecospheres' by scooping up and sealing seawater and sediment in glass bottles of one or two litres. Bacteria and algae in a mixture of air and saltwater were maintained only by sunlight and by 'doing their "microbial thing"', as Allen phrased it.[71] Folsome's bottled ecospheres remind of the half-pint milk bottles that Raymond Pearl had used to breed populations in the 1920s. In contrast to Pearl's fruit fly populations, however, Folsome's ecospheres seemed to 'survive' forever. It was their 'diversity' that kept their contents stable or 'alive'.

If the earth's biosphere could be considered a closed ecosystem, the ecosphere could be considered its simple model. Space Biospheres Ventures addressed Folsome as well as the Russian scientists in the search for concepts of how to overcome the limitations that had been faced in the existing experimental life support systems of space flight. Allen called these systems 'truncated ecologies' or 'non-ecological systems', since they were either free of humans or based on a soil-less agriculture. Such systems, he argued, failed to close the material cycle and thus could not keep the biosphere in balance.[72] With Biosphere 2 he hoped to develop a full-fledged ecosphere from the available scientific data and ecological experiences. The biosphere that Allen and his colleagues put into practice followed the same line as Vladimir Vernadsky's biosphere theory that they became familiar with in 1970 when G. Evelyn Hutchinson published his seminal issue on *The Biosphere*.[73] In cooperation with the American microbiologist and ecologist Lynn Margulis, a close collaborator of James Lovelock, the biospherians combined Vernadsky's work with Lovelock's Gaia theory and elaborated the concept of the biosphere into an evolving homeostatic feedback system. Jointly they invented a complex superorganism that would follow its own evolutionary interests and provide its own source of energy, as Allen explains:[74]

> Lovelock and Margulis showed in detail how the biosphere could operate as a *cyber-netic* system – *cyber* meaning helmsman – a system which is self-regulating by means of rapid microbial response to small changes in the composition of the atmosphere.[75]

In his account of the history of ecosystem ecology, Joel Hagen points to two related ideas central to Lovelock's Gaia theory. First, 'the biosphere is self-reg-ulating'; second, this self-regulation can be seen as 'analogous to homeostatic mechanisms in organisms and cybernetic controls in automated machines'.[76] Gaian concepts of a holistic earth as a living regenerative organism never opposed a cybernetic view of earth as a self-sustaining system. It was this analogy of the holistic and the cybernetic approach to life that the biospherians put to work. Next to Vernadsky and Lovelock, Allen referred this work back to the broth-ers Howard and Eugene Odum whom he called the 'total systems ecologists'.[77] The view that 'the biosphere is a homeostatic or cybernetic system of living and nonliving components', Hagen elaborates, 'was a central feature of the ecosystem concept that [Eugene] Odum did so much to create during the decades follow-ing World War II'.[78]

'The cybernetic process does not have to be thought of in terms of willful or conscious decision-making', Allen explains.[79] The cybernetic structure of Biosphere 2 fell victim to the same paradox as Spaceship Earth: Both were hier-archically organized and envisioned the expert scientist or engineer figuring as the helmsman on the bridge. And both accentuated principles of self-regulation: 'Life, the biosphere, creates its own controls and balances'.[80] Eduard Sueß's origi-nal concept of the earth's biosphere as the sphere encompassing life on earth was explicitly revised. The biospherians argued that the term biosphere, although increasingly used, had been 'inadequately defined by many as the thin layer of life on the surface of Earth'.[81] They challenged the often voiced assumption of the earth as vulnerable and dependent, and 'of life as fragile, a helpless passenger on "Spaceship Earth"'.[82] Instead, the disciplines of 'biospherics' and 'ecotechnics' would assist in studying the earth's biosphere as a powerful geologic force and in putting its laws into operation in a materially closed and informationally and energetically open system. Biosphere 2 would be modelled on 'Biosphere 1', as the project managers called the living earth, and would provide an upgraded biospheric operating system that was itself perceived as 'a cybernetic system moderating environmental conditions on our planet for the continuance and spread of life', stable, complex, adaptive and evolving – in short, a 'Life-Enhanc-ing Feedback System'.[83]

The 'technosphere' denoted the technological infrastructure, administrative organization and monitoring systems to sustain the distinctive eco-environments or biomes in Biosphere 2. The technosphere, as it were, had 'to stand in' for those geological functions of the planet that, like weather and climate systems, could not be ecologically modelled and performed on the mesoscale.[84] A 'sub-story' to the biospheric plot, a heavy-duty concrete fundament was erected that encased 'a lower basement-level story devoted solely to technosphere systems facilitating water, air circulation, and heating and cooling equipment'.[85] Within the technosphere, ecological relations were translated into biological, physical and chemical causes and effects. An elaborate structure of electrical, mechanical, chemical, thermal and hydraulic transmissions formed the basis of the nature that stretched across its surface like a thin organic skin. The underground technology powered the ventilators, pumps and turbines responsible for circulating, flushing, cleaning and tempering water and air. Technology also ran the sprinklers, the misting machines, the wave engine, the desalination device and other contraptions on the biosphere's surface.

The surface machinery was hidden from sight, as was the intricate network of more than 2,000 sensors that constituted the 'nerve system' of the plant. The sensors provided for continuous monitoring and communicating of vital features like air pressure, temperature, humidity and precipitation. The sensors also served to control chemical compositions and concentrations of oxygen, carbon dioxide and trace gases to keep the system stable. Information and data were analyzed in a laboratory in real-time for automatic or manual regulation, and then stored in a computer archive that could be assessed at any time from the inside, but also by Mission Control outside of the glass dome, the 'nerve center' that was also in charge of maintaining 'interbiospheric communication'.[86]

Glass Ark

Vernadsky's biosphere theory had described a self-regulating system that required all biotic and abiotic elements to carry out certain functions in order to keep dynamic equilibrium. The theory also implied that, in principle, elements could be replaced as long as their substitutes fulfilled the same basic biogeochemical functions. This systemic approach towards the necessary and the practicable materialized when building and populating Biosphere 2. Principles of the functional and the modular were implemented when constructing the spaceframes and also when assembling the biomes to constitute the mesocosm. Not all of earth became part of the second biosphere. Polar regions for example were not modelled. As their main purpose on earth was considered to be climate regulation, they were substituted by technology.

Modular principles reigned also on the meso- and microscale of marshland, rainforest, savannah, desert and ocean biomes. Artificially built components, like the big mountain and rock formations in the tropical forest, were combined with 'chunks' of nature hunted and gathered, or simply donated by other institutions. From all over the world, animals, plants, sands and soils were transported to the construction site in order to form a highly efficient concentrate of nature.[87] The indoor marsh and ocean system presented the largest artificial system of its kind ever created. This system, Allen explained, had 'to be composed only of high yield components: beach, lagoon, coral reef and adjacent ocean in order to compensate for its relative lack of area compared to Biosphere 1's ocean'.[88] The basement housed a mechanical algae scrubber system. To assemble the visible ocean and marshland, an entire living coral reef was brought in from the Caribbean Sea. Mangrove trees were shipped from Florida. Finally, thousands of gallons of ocean water were hauled in from the Californian Pacific coast. The greater part of the ocean came into existence by mixing up a fresh water salt recipe called 'Instant Ocean'.[89]

Similar to the wilderness biomes, the 'Intensive Agriculture Biome' obeyed the objective of 'maximizing the yield'. Soils and plants finely tuned to one another allowed for close crop rotation cycles. The biospherians expected to seed, cultivate and harvest as effectively as possible on a small surface, in line with future plans of extending their biosphere to other planets. 'On the Moon or Mars, with space at a premium, the ability to produce great amounts of food in small plots would be desirable'.[90] As important as sustaining maximum yields was maintaining closed material cycles. Again, this ideal involved the installation of technology, like quick composting machinery, in the invisible basement. It also required providing for the complete recycling of wastewater and sewage for irrigation and manure.

Allen drew on the motif of the ship to underline the need for unity and community in the first and second biospheres, emphasizing that 'spaceship Earth ... was a boat we were all in together.'[91] At the same time he made clear that not all of earth's inhabitants were meant to register for the journey. In a functionally designed mesocosm of limited space, the components would have to fit together in an optimal way. The 'Glass Ark'[92] emphasized not encyclopaedic completeness but systemic integrity. Not natural affluence but systemic 'diversity' was modelled, based on biological agents selected according to criteria of efficiency, practicality and replaceability.[93] In resonance with Liebig's Law of the Minimum, Allen cites Buckminster Fuller who allegedly held, 'We all know the saying that a chain is only as strong as its weakest link'.[94] What did this entail for the proverbial earthly ship?

Legend tells that Noah admitted each kind of animal, two by two, to his ark which sheltered them until the deluge had passed. Details of diet and cooperation between the animals during their stay in the ark, however, remain a mystery enshrouded by time and myth.[95]

The animals of Biosphere 2, along with the soils and plants, were chosen in a strict 'selection process',[96] which did not adhere to the biblical principles of taking two of each kind:

Unlike the ark, Biosphere 2 welcomed animals to a *web of life*, as participants in the *oikos logos*. From the Greek origins *oikos* (house) and *logos* (governing rules), ecology literally means the 'rules of the house'.[97]

'The rules of the house for Biosphere 2', Allen continues, 'were much the same as for ecosystems of the Earth'. The biospherians believed in the ecological truth that '[l]ife is an ecological property, and an individual property for only a fleeting moment'.[98] As addressed in the previous section, they recognized and valued not the individual living being so much as the ecological community. Diet among the animals was assessed in the larger evolutionary scheme: 'To eat and be eaten is the fate of all creatures'.[99] Cooperation between the animals stressed the functional qualities of life within the ecological unit.

Residents had to earn their keep, performing some useful function in the ecosystem. All the work of the ecosystem had to get done – all the functional niches must be filled ... Biosphere 2 was similar to an island ecosystem in that the residents could not leave, and new populations or individuals could not migrate in to rejuvenate or replace resident populations.[100]

'Noah had it easy compared to this!'[101] In a tedious process, about 3,800 species were taken on board. Not the biblical disposition of clean and unclean but usefulness, size and composure made the difference. 'All the animals selected are gentle and easy to handle'.[102] Similar to the space freighter *Valley Forge* in the movie *Silent Running*, the chosen animals and plants were small and peaceful. They were moderate consumers. Dangerous or poisonous animals and plants were forbidden. 'The only large predators allowed inside the Glass Ark are humans'.[103] Linnea Gentry and Karen Liptak report on the example of introducing the hummingbird to the biosphere, which required a choice among the roughly 340 known species of hummingbirds. 'The final candidate had to fit a very demanding list of qualifications'.[104] Like a special tool for a job the specimen had to meet precisely circumscribed criteria to perform the specific task of pollination: 'For Biosphere 2's wide range of flowers, a hummer with a straight beak about 20 millimeters (8/10 of an inch) long was required. The pretty sapphire-spangled hummingbird fit the bill'.[105]

Species conservation was not the concern of Biosphere 2. The main purpose was maintaining overall stability, which meant to avoid sudden population growth and decline. 'Expansion is designed to occur modularly, in discreet biospheric units, not by risking "overstocking in the petri dish" with the resulting crash of key populations'.[106] The ecological equilibrium asked for a rigorous control also of the human population, which had to be balanced at its optimum number.

> Most critically, in Biosphere I both humanity and its industrial works are exponentially increasing while Biosphere II is designed for a maximum human population of ten, a working population of eight, and has its industrial/laboratory areas restricted to definite volume.[107]

When four women and four men moved into the biosphere for the two-year closure experiment, the balance of the sexes was in line with the overall schemes of symmetry and equilibrium in Biosphere 2. The configuration also mirrored the biological ideas of reproduction essential to the project, even though human reproductive stability was not addressed. The project managers were not interested in researching models of utopian societies in space. The media closely attended the experiment and speculations abounded about how human cooperation would play out in the enclosure. Under the header of 'Keeping Healthy', Allen quoted the daily newspaper *Washington Post* in asking, 'Will there be sex in the Biosphere? Of course, but who cares … '[108] Numerous scholars interested in psychological, sociological and anthropological studies approached the project team in the preparation of Mission One. The team, however, decided to focus attention on ecological ecosystem functions and on questions of the natural sciences. 'The private journals of the crew may someday reveal some interesting aspects of the dynamics of isolated colonies that will be applicable in the new world of space exploration'.[109]

The human beings introduced to Biosphere 2 also had to meet specific criteria in order to function as useful agents in the biospheric metabolism. Like all other living beings entering Biosphere 2 humans had to undergo a tough selection process. Most of all they had to be physically healthy and athletic. They were educated scientists and engineers trained to perform a multitude of extra tasks, from diving to gardening to butchering. Resembling the scientist-astronauts entering US space stations, the biospherians were physicians, physicists, biologists, ecologists or electrical engineers. They had to be 'self-sufficient in everything from running tests in the labs to doing their own repairs.' They were 'their own plumbers, electricians, and tailors, too … They've been trained to be good observers and researchers, as well as competent farmers and computer users'.[110]

In 1991 eight biospherians entered the enclosure in the familiar way astronauts enter their spacecraft: in orange jumpsuits and intensely covered by the
media. Their number materialized the concept of the optimum population on
the mesoscale. The working population of eight combined utilitarian principles
with a notion of community: 'Eight people appeared sufficient to handle the
workload necessary for such a venture. Eight were numerous enough to keep
each other company'.[111] The number eight was also indebted to ambitions of
space exploration:

> Eight was the original number for the crew of Space Station *Freedom*. The number
> eight had also been suggested as ideal for an international manned mission to Mars.
> And eight had long been the size of the basic unit of military life; a squad consists of
> eight soldiers.[112]

The notion of 'crew' emphasizes the spaceship journey, while 'squad' highlights
the military aspect of the endeavour and the 'basic unit' stresses its pioneering
zeal. Jointly they seem to confirm Sloterdijk's proposition of the selectiveness of
the ship's voyage. The chosen ones form a singular heroic elite that may prove
fierce and militant, pioneering and groundbreaking, or simply tragic and heart-
breaking. The chosen ones may be doomed, or rescued, or both.

The Economy of Replacement

In 1929 John Desmond Bernal predicted that globes would one day replace the
earth. In detail he outlined how a spaceship or a permanent space station – a
'globe' or 'colony' – would have to be constructed to take over the earth's vital
functions. As if in anticipation of Biosphere 2, he projected an artificial sphere
that 'would fulfill all functions by which our earth manages to support life.' Bernal also delineated measures for continuous material recycling, 'for it must be
remembered that the globe takes the place of the whole earth and not of any part
of it'.[113] We could easily agree with Gentry and Liptak, who, in a slight modification of Buckminster Fuller's line about all of us being astronauts, maintained
that, 'Actually, we are all biospherians'.[114]

Biosphere 2 may be the most prominent project ever undertaken to frame
the earth in ecosystems terminology and to merge it with a veritable spaceship
architecture in order to construct a high-tech habitat elsewhere that reversed
the relations of biosphere and technosphere. In 1994 the eco-visionary Kevin
Kelly marvelled, 'life is a technology. Life is the ultimate technology'.[115] Science
author Dorion Sagan, son of Lynn Margulis, hailed the Second Nature that was
created in Biosphere 2.[116] The second nature formed a part of the world and yet
was separated, held in store until it would serve as technological substitution
of the first.[117] It enrolled life and living space into a new economy of techni-

cal maintenance and expert control. Within such economies of nature, social theorist Timothy Luke argued, a nature formerly accessible and free of charge might turn into an exclusive and costly resource.[118] The commons of oceans and atmosphere, and the climate processes they uphold, will be subjected to new negotiations over access and distribution. In 1990, when Biosphere 2 was about to be closed for its first mission, the science fiction movie *Total Recall* premiered to illustrate how within a hundred years the allocation of biospheric supplies enable and constrain life in a colony on Mars. The plot revolves around a corrupt governor who monopolizes the production and circulation of air to control rebellious groups demanding independence. The movie exemplifies the social and spatial stratification that may result in enclosures where access to life support is administered according to practicability, utility and good conduct, and controlled by economic, political and technoscientific power.[119]

Biospheres make excellent cases to illustrate the politics involved in natures and environments. They form perfect examples of the effective moral economies. And they demonstrate that substitute natures redefine the problems which they had been sought to alleviate. Second natures create second problems. Notably, in Biosphere 2 neither acid rain nor overpopulation nor pollution posed a problem within the artificial environment. Instead, harvest failures and food shortages, a decline in pollinating insects, losses of vertebrates and other species and the explosive proliferation of ants complicated life under glass.[120] William Mitsch, editor-in-chief of *Ecological Engineering*, acknowledged that 'CO_2 concentrations soared and O_2 concentrations dipped during those first years', in a preface to a special issue on Biosphere 2 in 1999.[121] Though 'heavily subsidized', the closed material cycle could not be made to work, not even at all costs.[122] Like so many myths, the story of the Second Biosphere met its fate in hubris. The *endosphere* turned out to be an *exosphere*. The only environment to live in was *outside*.

Speaking of hubris in relation to Biosphere 2 is not without controversy. The builders of Biosphere 2 go as far as acknowledging a loss of certain species and an oxygen imbalance that called for an injection one and a half years into the experiment. 'Noah's ark is a metaphor, not a schedule', acknowledged Peter Warshall, Biosphere 2's savannah ecosystem designer; 'restoration or construction of a community cannot be instant, but requires patience and attention'.[123] Yet, when Mission One ended in 1993, biospherians Alling and Nelson maintained that 'At the end of its two-year maiden voyage, Biosphere 2 is like a well-tested ship, ready to carry on research in biospherics for decades to come.'[124] In contrast, media coverage turned from sensationalist to critical to sneering.[125] Nelson and Bill Dempster, the director of systems engineering until 1994, deplored those media stories that 'fantastically exaggerated' the ambitions of the project and 'later attacked furiously when the magical expectations were not met'.[126]

What interests me is not whether the project worked or failed. I am interested in the dominating images of human-natural relations the project promoted. The experiment's provocation consisted in the proposition to understand the environment not as embedded, permeating human affairs and humans, but to perceive it as an enclosed and external object of potential human management. 'The real point that resonates with me about Biosphere 2', Mitsch contemplated, 'is the sheer magnitude of what it costs in money, material, and energy to create enclosed healthy ecosystems, not a trivial point if we are some day interested in habitation in space'.[127] Biosphere 2 offered a novel field of research and new insights into the works of nature, which turned out a wealth of peer-reviewed results.[128] Nevertheless, in 1999 scholars of ecology, among them the pioneering ecosystems scientist Howard Odum, admitted that the 'experiment thus far has shown the difficulty in recreating the viability of our planet, Biosphere 1. Scale and an over-packed inventory of the familiar plants and animals that we know, may not be enough or may require a longer time to more resemble planet earth'.[129]

Luke chose the term 'denature' to describe the nature of Biosphere 2, a nature that offered space neither for emptiness nor for affluence. Unlike Second Nature, denature indicates that the technological environment of Biosphere 2 represented a simulation for which an 'original' did not exist.[130] This is not to say the world of Biosphere 2 was lifeless. Most of the inhabitants did actually 'survive'. The term denature rather focuses on the aspirations of the experiment to build a stabilized world, an equilibrium that had to exclude all natural traces of chance, disorder, difference, sickness and other forms of disturbance. The technologically sustained and monitored environment appeared lush, fertile and thriving, yet it was not unlike conventional minimalist ideas of life support in space. 'As with the Apollo-13 flight to the Moon', Eugene Odum mused, 'survival becomes the mission when life-support is in question'.[131] To biospherian Jane Poynter the dome was a glass 'cage' in which she and her fellow crewmembers were 'inmates'.[132] The French philosopher Jean Baudrillard addressed Biosphere 2 as a 'glass coffin', thereby expressing his criticism of conserving and musealizing nature during its lifetime. Following Baudrillard Biosphere 2 aspired to immortality by adhering to a minimalist principle of 'survival' through steady regenerative succession – 'just like in paradise'. Baudrillard suggests that the experiment centred on 'dissolving, shrinking the metaphor of the living into the metastasis of surviving', the reduced continuation of life that avoids death.[133]

Nature in a State of Exception

Logan's Run featured a stabilized community, captured under a dome and controlled by a computer system to which the inhabitants are perfectly oblivious. Their closed world appears as a generous, liberal and thriving world. Logan's job is to hunt down those who doubt the scheme and try to escape from their certain premature death. Ultimately he finds himself on the run. He struggles to find a place called Sanctuary, following 'a pre-catastrophe code word used for a place of immunity'. Sanctuary turns out to be an unidentified place of wilderness beyond the seals of the domed city and beyond the reach of the authorities. The central processing unit has not been programmed to comprehend the meaning of either Sanctuary or *Ankh*, the old Egyptian hieroglyphic symbol for life that serves Logan as a key to freedom.

The forgotten outside world that Logan and his partner Jessica stumble upon stands in stark contrast to the polluted and devastated earth one would expect. While nature was left untouched over a course of three centuries, it regenerated and grew back into a lush and beautiful jungle of thick forests and tall grass. Logan and Jessica meet an abundance of birds and animals, clear water and a perfect sunrise. They do not immediately realize that this is the refuge they were looking for. Only gradually they understand that their sanctuary is neither sheltered nor controlled, neither stabilized nor enhanced, but opulent and entirely unbound.[134]

The end of Biosphere 2 came about even faster than the end of the domed city in *Logan's Run* and the end of the space freighter *Valley Forge* in *Silent Running*. In all three cases, political resolutions and revolutions ended the projects way ahead of their scheduled time. Shortly after the resealing of the glass dome for its second closure period in 1994, a conflict broke out among the operators and the second crew of Biosphere 2. The dispute was taken to court and the mission already under way was aborted. In 1996 the facility was placed under the management of Columbia University in New York. Since then Biosphere 2 has served as a greenhouse for research and education.[135] In the movie *Silent Running* decisions made at the base on earth lead to the mission's termination. Following a short battle and a long solitude, Freeman Lowell, the sole human survivor aboard, blows up his spaceship, but not without arranging for a last bubble of life to survive floating in space. Significantly, this ark is devoid of humans: a small robot takes devoted care of the dispatched idyll of plants and animals – a suggestion perhaps that earth does not need humans to continue to exist.[136]

It is not without irony that Biosphere 1, the earthly environment that was more and more considered an obsolete version of nature, had to stand in as a reserve for Biosphere 2. In the form of the spaceship, the ark and the frontier outpost, Biosphere 2 acted as a heterotopia, as a mirror, inversion or continuation of primary nature.[137] Sites like Biosphere 2 generate extreme or exceptional situations by transforming the demarcated reserve, the nature to be protected from human access and use, into a reserve or reservoir for newly created environments. The nature reserve comes attached with exclusive reservations and rights of access that determine who in case of emergency will be able to command, to use and to dispose of this nature.

Technonatures like the *Valley Forge*, the Mars colony in *Total Recall*, the sealed city in *Logan's Run* and also Biosphere 2 have carried to extremes the image of nature in a 'state of exception'.[138] The Italian philosopher Giorgio Agamben suggested that the sovereign power of the modern nation-state is founded on a conception of sacred or 'bare life' that in a state of exception can be exposed, excluded, exploited and even eliminated, but not sacrificed. Following Bruno Latour's observation of how, according to the 'modern constitution', nature has been separated from culture, modern ecology and ecosystems sciences have constituted the condition of nature by a notion of the exceptionable.[139] This paradox was also operative in Biosphere 2. The second half of the twentieth century saw a system of ecological governance that claimed and exerted control over nature and, facing a situation of ecological emergency, authorized nature's infinite detention, exploitation and enhancement. "'We are as gods," Stewart Brand, environmentalist and editor of the *Whole Earth Catalog*, had written in the catalogue's 1980 edition, "so we might as well get good at it."'[140]

6 DEPARTURE: THE HABITATS OF TOMORROW

Zed: This place is against life. It must die.
It's a prison. A prison.

Friend: No, it's an ark.
A ship. A space ship.
All this technology was for travel to the distant stars. – Mm.

Zed: Did you go?

Friend: Yes. Another dead end. [chuckles]

The Vortex

More than three hundred years into the future, the concerns of the 1970s – population growth, resource exploitation and environmental devastation – are a matter of the past. In 2293 the world is divided into two separate societies. The Vortices encompass the few isolated and luxuriant refuge communities of immortal intellectuals, artists and scientists. Outside of these remaining civilizations, the savages live in a bleak world of scarcity, pollution, war and death. The Outlands are habitually raided and plundered by the Brutals, a horde of primitives who kill in the name of their fearsome god, Zardoz.

The movie *Zardoz* was released in 1974.[1] From the onset the spectator is aware that Zardoz is a fake god invented by a clever Vortex 'puppet master' in order to gain power over the Brutals and control the Outlands. The name Zardoz is an abbreviation of the long-forgotten American classic *The Wizard of Oz*. In the book, a weak human being hides behind an elaborate masquerade and technical apparatus to hold power over the land and its inhabitants. In the world of *Zardoz*, the troops force the population into cultivating the land and then seize the yield gained by primitive plowing and farming to increase the god's wealth. The Brutals are also agents of extermination. Their deadly raids are intended to keep the population of the Outlands balanced; their killings prevent the people from creating 'new life to poison the Earth with the plague of men as once it was'.

Sean Connery plays the Brutal Zed who stows away to gain access to the Vortex and reveal the truth about the godly ploy. Zed manages to enter the small but rich community of Vortex 4. Colourful, youthful and peaceful, Vortex 4 provides a healthy and mainly self-sufficient environment. 'Welcome to paradise', Zed is greeted upon entering this oasis which cultivates traditional techniques of farming, plant breeding and animal domestication, corn grinding, baking, spinning and weaving. The local subsistence economy is premodern in its daily routines if it were not for its high-tech gadgets. The combination of the ancient and the futuristic is a characteristic feature of science fiction stories about survival communities surrounded by hostile environments. Vortex 4 exchanges goods with other vortices according to its needs and surpluses in a traditional barter economy run by means of high-tech communication and information technologies. Vortex 4 relies profoundly on modern technology and science, on genetics, mathematics, physics, chemistry and psychology.

Welcome to paradise, indeed. The society of Vortex 4 is one of 'absolute equality', a perfect democracy, as Consuella, one of the strong female figures in the community, is keen to assert. All concerns are subjected to community decision according to majority vote. A judicial system and police force are unnecessary; information and surveillance technologies suffuse all of society and see to instant discipline and punishment, not by imprisonment but by aging. 'Let's take yet another boringly democratic vote, shall we, Consuella?' Friend, a Vortex resident, expresses the boredom bred by two hundred years of the same monotonous life in perfect 'equilibrium', a life without surprise, disruption, change or objective: Friend acts as the informant and the cynic provocateur; he illustrates the weariness and the apathy of a society in which individuals live forever.

The Vortex in its eternal bliss has become a sterile place. The 'delicate balance' of the 'perfectly stabilized' community requires neither death nor birth; residents have long discarded sexuality, preferring infertility to sexual reproduction and procreation. The listless Apathetics have long succumbed to the meaninglessness of a life of personal indifference. The aged Renegades have stubbornly defied the strict societal rules and have been condemned to a life of eternal senility. The Renegades are the nonconformist memory keepers of pre-apocalyptic times, like Thorn's companion Sol in *Soylent Green* and Freeman Lowell in *Silent Running*. The Renegades also serve as markers of time in a timeless society in which death has been banished. In a case of accidental death a person is reconstructed at once, as the Fathers resolved three hundred years ago: 'Here, man and the sum of his knowledge will never die, but go forward to perfection.' Friend passes the origin story of the Vortex on to Zed: 'They were the scientists, the best in the world.

But they were middle-aged, too conditioned to mortality. They went Renegade. We were their offspring, and we were born into Vortex life.'

The Vortex serves as a vast archive, as a library, a conservatory and a museum, storing the classic works of art in order to perpetuate the traditions and cultural heritage of former civilizations. Knowledge has been preserved and cultivated, from history, natural history and anthropology to philosophy, music, the arts, languages and literature. The commune also stockpiled crafts and artwork from antique Greek statues and ancient books to outmoded clocks and cars to modern paintings. The Vortex was planned as a sanctuary and as a cradle of civilization at the moment of decline. When Zed is hunted down as an intruder to the austere community he begins to comprehend the extent of the conservationist mission that withered the present in an attempt to safeguard the ancient past. The inhabitants have become hostages of their own laws passed to rule out difference, disorder, sickness and death. The Vortex sustains not life but the conditions for survival, a highly sophisticated form of life support. Ultimately, however, the Vortex lives up to its name; it constitutes the core of knowledge and also the swirl of obliteration. Ironically, it is not life but oblivion that the community members desperately crave.

In time the community perceives Zed as the Chosen One who will relieve the Vortex population of the curse of eternal life. The movie *Zardoz* is saturated with references to religious belief as a means of violation, control and liberation of life and living beings. Religious themes recur as the movie negotiates mortality and immortality, proliferation and obliteration. Last but not least, religious motifs are present in the story of Zed and Consuella surviving the destruction of the commune and becoming First Man and First Woman. As Adam and Eve they embark on a new journey, finding comfort not in the promise of eternal life but in the eternal cycle of life.

Homo Spaciens

Looking at the second half of the twentieth century and how Western science and technology addressed the earthly environment, assessed its inventory, allocated resources and gauged the costs of containing the human environment in reserves, arks, lifeboats and spaceships, one could reasonably assume 'someone may soon contend that ecology boils down to space flight':

> Perhaps a direct path leads from the 'planetary' experience of *Sputnik* via Sänger's Malthusianism and Boulding's 'Spaceship Earth' to the 'ecological' problem of overpopulation brought up by the Club of Rome, and from there to the Greens on the one hand, and via *Global 2000* [an environmental study issued by U.S. President Jimmy Carter in 1977 and published in 1980, S.H.] back to the American President on the other, whose name by this time was no longer Kennedy but Reagan, and who

sent NASA to its 'second stage'. Closing the divide, someone may soon contend that ecology boils down to space flight.[2]

It may be arguable whether the unprecedented conjunction of ecology, geopolitics and technoscience in the twentieth century was directed towards leaving the earth, as Michael Jäger and Gudrun Kohn-Waechter presume. Unquestionably however, associations of environmental degradation and the human flight into space have become familiar and frequent in the last decades. Following Jäger and Kohn-Waechter, these relations could be termed 'strategic' in the sense of the 'great anonymous, almost unspoken strategies' that Michel Foucault pointed to, collective rationalities and connections that are seldom directly expressed and yet are purposeful and omnipresent.[3]

The inventors of Biosphere 2 were clearer than others. They considered going to space and expanding Earthlife into and beyond the solar system a 'historical imperative'.[4] The lifespan of Biosphere 1, the project managers warned, was at best equal to the life of the sun. More probable yet, the life of the earth would be shortened further by sun changes, cosmic impacts and by the failure of ecological homeostasis and an ensuing entropic reaction. As a likely reason for the death of the earth, the biospherians considered the abuse of the atomic bomb and the environmental catastrophe of a nuclear winter. In addition, they pointed to the human-induced 'burdens of population explosion, agricultural stress on the environment, technological paralysis, mineral and fossil fuel depletion, and a number of other potentially debilitating conditions of an overburdened planet'.[5]

Nonetheless, the founders of Biosphere 2 regarded the second half of the twentieth century not only as the apogee of doom but also as a singular 'window of opportunity', as the short historical time span in which humankind had acquired the means to turn itself into a spacefaring civilization. The features that characterized this exceptional window were a serious concern for the environment, an unparalleled concentration of wealth, resources and energy, and, lastly, the adequate technologies at hand. Before too long, the project leaders cautioned, this window would close again as environmental burdens would prevent the expansion of the human species into the solar system. The first biosphere would be 'doomed to die within a small fraction of the life of the cosmos'.[6] The project of creating a second biosphere followed a theory of biospheric evolution, of biospheres being able to evolve and survive instead of dying off in a Darwinian fashion: 'Biosphere I must disappear sooner or later unless it can participate in sending forth offspring biospheres to populate other regions of the cosmos'.[7]

Biosphere 2 did evolve, but it never took off. After the two periods of closure, from 1994 onwards, Biosphere 2 was operated as an 'open system', with a research focus on ecological processes within its seven biomes. Biosphere 2, though at a scale exceptional in experimental ecosystem research, became one of many facilities to study the ecology of Biosphere 1. All the while, aspirations of humans leaving the earth in artificial biospheres did not subside, but they took new forms. In 1982 a Spaceship Earth was inaugurated in Florida's *Walt Disney World*. The spherical structure followed Richard Buckminster Fuller's geodesic dome design, and its eighteen-floor interior was furnished by a geosphere built to transport visitors across the farthest reaches of time and space. The view from the outside was that of a shiny silver spaceship of more than fifty meters in height and resting on a tripod as if ready for take-off. Disney World's Spaceship Earth still forms a central part of the theme park EPCOT, the acronym for Experimental Prototype Community of Tomorrow.

In the summer of 2006 the renowned British physicist Stephen Hawking revitalized the public discussion about colonizing space by posing a question in a *Yahoo!* Internet forum: 'How can the human race survive the next hundred years? In a world that is in chaos politically, socially and environmentally, how can the human race sustain another 100 years?'[8] Addressing the future of mankind facing viruses, terror and war, along with resource exploitation and global climate change, Hawking answered his question by suggesting that survival could in the long run be secured only by humans swarming out into space and colonizing other planets. By manipulating their genetic material humans could be bio-engineered to make them capable of tolerating the environments on other planets. A new species adapted to life elsewhere in the galaxy could be created – *homo sapiens* developing into *homo spaciens*[9].

Suspicions of an ecologically motivated abandonment of the earth such as the misgivings voiced by Jäger and Kohn-Waechter, or forthright suggestions like those of Hawking prompt one to reflect on how visions of Spaceship Earth may generate strategies neither intended nor sought by ecologists. The view of earth from space does not necessarily lead to environmentally correct conduct. Whole Earth, Hans Blumenberg noticed, has not been turned into humankind's adopted homeland. The earth seen from space did not even include humans, neither their creations nor their waste.[10] Just as often, space flight seems to be viewed as an exit strategy – a means for escaping an environment humans have worn out. In 1980 the popular American cartoonist Gary Larson published a cartoon image in which animals, two of each kind, move towards a NASA spaceship as their salvaging ark.[11] While the animals clearly sense the impending catastrophe, the humans in the person of a baffled gatekeeper appear to be clueless. Their elaborate technoscientific apparatus of monitoring and control has failed; *homo sapiens* (or is it *homo spaciens*?) does not have the imminent ecocide onscreen. The image inverts the power relations in the biblical story of the ark, reversing the Christian teleology of the species selected for survival being the one made in God's image. Also, Larson gives a modern and ironic twist to the story of the ark by having the animals choose as their lifeboat a spaceship, the epitome of human technologies designed to achieve world control.

The laugh is on us. Larson's joke reveals that we are already part of its plot. 'The joke is a sign of successful interpellation', Donna Haraway proposes, 'of finding oneself constituted as a subject of knowledge and power in these precise regions of sociotechnical space. Whoever is inside that joke is inside the materialized narrative fields of technoscience, where better things for better living come to life'.[12] At the same time, Haraway reminds us that '"comic" means reconciled, in harmony, secure in the confidence of the restoration of the normal and noncontradictory. ... The comic does not recognize any contradictions that cannot be resolved, any tragedy or disaster that cannot be healed'.[13] We have become accomplices of the scheme the joke outlines. Enjoying Larson's space ark we disclose that we consciously and deliberately form a part of the human collective that prefers to act ecologically naïve, or cynical, or progressive. 'The comic mode in technoscience is reassuring in just this way'.[14] We insist on the continuing dependence on technoscientific solutions, forever hopeful that a way out will present itself.

Virtual Frontiers

In recent years, official space endeavours have focused not on 'manned' projects so much as on 'unmanned' explorations of distant planetary surfaces by robotic vehicles. These undertakings tend to legitimize the effort and the money they consume for exploring the universe by highlighting the by-products of space research. Among such increasingly important spin-offs is the accumulation of data through close monitoring of planet earth for environmental information gathering. In the new millennium, space flight and ecology have settled into a strategic alliance that political scientist Karen Litfin has called the 'global gaze'.[15]

When Google launched the project 'Lunar X PRIZE' in 2007, the goal of saving the earth by going into space was one of the major aspirations.[16] The $30 million international prize was broadly advertised with the claim that 'it is our duty to use the resources of space to help our fellow passengers on spaceship Earth.'[17] The promotion of the prize called up the familiar concerns and the hopes of the Environmental Age, but the actors, their priorities and their means had changed radically. Not a government but privately funded teams prepared to make history. The prize did not reward personal cut-downs for the sake of environmental quality but stipulated new quality technologies of earth exploration and observation brought into space by profit-based private enterprise. The information gathered in the project eventually enabled a new visual tool, the Internet service Google Moon, launched in 2009 for the fortieth anniversary of the Apollo 11 moon landing.

The fifth version of Google Earth, unveiled by Nobel Prize winner Al Gore in February 2009, also featured an application called Google Mars. To explore the surface of the Red Planet Google Mars used the latest high-resolution imagery and data in collaboration with NASA scientists.[18] To paraphrase John F. Kennedy, what was once the furthest outpost on the old frontier of science and space became the furthest outpost on the new frontier of virtual space. Remote sensing technology, visual data processing and imaging tools claim to convey close-up views of virtually every spot on the earth and beyond.[19] These versatile technologies replaced the Whole-Earth imagery of the 1960s and 1970s as a visual expression of Spaceship Earth.[20] Internet applications and commercial platforms like Google are praised for raising environmental awareness by monitoring environmental alterations like deforestation, melting polar icecaps and rapid urban sprawl in real time.[21]

At times, just as the figure of the spaceship did, earth monitoring inspires expert global players to manage the planet at liberty and will. The game SimEarth, for example, addresses players of the information technology-saturated Western world who enjoy the popular Sim franchise but find that a city is not large enough to satisfy their creative needs. They are given the opportunity of a simulator lending full control of an entire planet.[22] The game provides users with a push-button position to craft and operate their own earths. The earth operator runs the geosphere, directs the evolution of microbes into intelligent life, adds vegetation, populates the planet with animals and humans and builds up civilizations, a process that effortlessly covers time spans of billions of years. By choosing 'New Planet' from the menu, earthly problems like pollution, disease, famine, war and global climate change can be sidestepped. The player can start from scratch and rerun the program whenever so desired.

Correspondingly, Sim-Bio 2 is a computer game to simulate the metabolic processes in Biosphere 2. The game uses data from the actual Biosphere 2 experiments and feeds them into a default model of Biosphere 2 that can be accessed or newly assembled by selecting and connecting components from a library of Biosphere 2 variables. The resulting model can then be run to simulate experimental results.[23] Simulations like SimEarth and simulations of simulations like Sim-Bio 2 create second natures, third natures and technonatures of higher orders that in contrast to Biosphere 2 will forgive all environmental blunders and failures. Haraway has pointed to a fundamental misunderstanding of Lovelock's theory of Gaia that enables just this kind of technoscientific recreation. Lovelock, she argues, never considered the living tissue of the planet as 'organic, nontechnological, alternative' nature. 'Quite the opposite, the view of terran life from a satellite or a spaceship is semiotically, but also technically, intrinsic to the Gaia hypothesis.' It is exactly the technoscientific nature of Gaia that is 'the one reason why Gaian thinking was built into the programming for the SimEarth computer game'.[24]

Whichever standpoint we decide to take towards the envisioned possibilities of the smart grids and self-configuring global sensing systems, questions and challenges like those brought forth by Hawking and answers provided by the new virtual and real worlds of earth and space exploration call new attention to an old theme: the alleged power of humanity to choose a technologically enhanced nature over a once-pure but now polluted environment by constructing a superior, more economic, robust and sustainable earth than the one that is now literally at disposal. At stake in these debates are not so much the moral issues of a 'natural' versus an 'artificial' environment and whether humans should abandon their home planet or stay and act responsibly. At stake are the consequences of humans' renewed confidence in their ability not only to pose these questions but also to resolve them.

The New Economy of Nature

When Hawking addressed the fate of the human race he used the verbs 'survive' and 'sustain', relying on principles of life support and survival that have become common and even fundamental in describing life in the twenty-first century. Life support has become the familiar term to denote the measures and equipment for the clinical stabilization of body functions in intensive care units. More generally, technical life-support systems come into operation whenever human life is to be sustained in abiotic environments in which life is considered impossible. In spaceflight, life support denotes the technical production and monitoring of a habitable cabin environment, regulating air quality, temperature, pressure and humidity, providing water and supplies, and disposing or recycling waste. In the ecosciences and environmental politics, a 'life support system' is defined as 'any natural or human-engineered (constructed or made) system that furthers the life of the biosphere in a sustainable fashion'.[25]

According to the *Encyclopedia of Life Support Systems* (*EOLSS*) that was developed under the auspices of UNESCO, the global concerns brought to the fore by the Earth Summit held in Rio de Janeiro in 1992 and its issued programme, the Agenda 21, widened the narrow meaning of life support systems in clinical practice to 'the whole of our planet as a grand intensive care unit'.[26] In 1987 the Ride Report on 'Leadership and America's Future in Space' advised the launch of an initiative called 'Mission to Planet Earth' to 'understand our home planet, how forces shape and affect its environment, how that environment is changing, and how those changes will affect us'.[27] In the same year the International Council for Science (ICSU) launched the International Geosphere-Biosphere Programme (IGBP) to research global change and organize data collection.[28] Together, these initiatives formed the beginning of what presently trades under the name of 'earth systems sciences'. In the tradition of the ecosystems sciences of the Environmental Age the loose international research network has set out to explore the inventory and behaviour of the 'earth system' by global scientific analysis and by modelling the interactions with human systems.

It seems nearly impossible to take a critical stance towards such sweepingly and compassionately formulated ideas and goals. The technocratic and experto-cratic undertones often go unnoticed. 'The oneness of the earth is understood according to this paradigm in system categories, its unity as the interaction of component parts, and the historical task as keeping the vital processes from destabilizing irretrievably', explains cultural and environmental historan Wolf-gang Sachs. Unity in the earth systems sciences is pursued 'through securing the necessary system requirements'.[29] Timothy Luke stresses the 'expertarchy' of earth system scientists that position themselves as earth managers and adopt a utilitarian view on nature as stock, service and system. In his view, the earth sys-tems sciences contribute to a notion of sustainability as a softer, gentler style of the longstanding 'sustained yield'.[30]

Concepts of optimum global space and surface capacities continue to be employed as a way of assessing earthly limits and constraints. In the 1990s the 'ecological footprint' became an obvious and meanwhile well-established meas-ure of environmental use and abuse. The footprint was directly derived from the ecological 'carrying capacity', defined by its authors as the maximum 'load' that humankind in the long run can put to the biosphere without risk.[31] The quantitative environmental assessments went hand in hand with economic ways of 'taking nature into account'.[32] In 2008, the American environmental scien-tist Gretchen Daily was rewarded Norway's Sophie Prize for her book *The New Economy of Nature*. The book harnesses not philanthropy or political legislation but an allegedly more potent force – self-interest – to preserve the earth's 'life-support systems'. The 'ecosystem services' of nature, Daily suggests, need to be priced high to create self-sustained markets for 'natural capital'.[33]

Ecological economists had balanced the books of nature in a similar way. In an oft-quoted *Nature* article from 1997, Robert Costanza and colleagues programmatically figured that the 'entire biosphere', that is, 'the services of eco-logical systems and the natural capital stocks that produce them', amount to the staggering economic value of an average U.S. \$33 trillion per year, most of which is outside the market.[34] Pointing out that adhering to the common prac-tice of externalizing the costs of environmental destruction will just increase environmental problems was a strategy to promote the emerging field of envi-ronmental accounting against the principles of the Western capitalist economy. But accounting for nature was itself expressing a view on nature as a utility that can be enrolled in accounting practices. The suggestion of internalizing envi-ronmental costs into the capital flow system for the sake of the environment affirmed the view of nature as an object that like any other good can be quantified, priced, commoditized and traded in market economies. This view carries the danger of seeing nature as an exchangeable object that can be substituted in monetary or in technical terms, or, if not profitable, simply discarded.

Travelling the World Without Leaving Home

The modes of global surveillance and management instigated in the Environmental Age continue into the present in earth satellite surveys, in planetary computer simulations and in enlisting nature into global markets. In some respects, however, the ideal of a universal vision as well as the ambitions to acquire such a privileged viewpoint have been forsaken. The 'Mission to Planet Earth' initiative of 1987 that concentrated its full gaze onto the earth also marked the beginning of the end of the technocratic and directed 'spaceship' as a meaningful earth metaphor. The Global Positioning System (GPS) brought into place by 1990, ready to be used for military purposes during the Second Gulf War (1990–1), was another such moment when images of the earth were literally projected back onto the earth instead of new planets.

The earth has again become a transient and fragile place to live. Human living space on earth is considered more limited and more threatened than ever. And yet Spaceship Earth and the idealized or dreaded habitats it offers have become unable to carry the weight of the world and its pressing issues of global change. The figure of a steerable earthcraft with its elaborate construction of a closed biosphere and a sustaining life support system has changed the notions of how truly luxurious human habitats should be equipped: expensive, shiny, secure, comfortable and air-conditioned – like the Renault Espace, one might add. But, like the Espace, these conditioned spaces no longer claim to contain and control, much less move the entire planet. The new sites are highly individualized, exclusive and withdrawn. They are conceptualized as remote, insulated sanctuaries amid increasingly wasted environments.

An exclusive space fashioned according to contemporary requirements is the 'seed ark' opened in 2008 in the permafrost of Spitsbergen, a Norwegian island in the Arctic Ocean. The *Svalbard Global Seed Vault* is a tomb structure built deep into a mountain of sandstone and ice. In the controlled underground atmosphere temperatures are held stable at lows of 0–1 degrees and heights of 40 degrees Fahrenheit to sustain plant biodiversity for future generations. Repeatedly referred to as a 'doomsday vault', the global seed bank offers every country in the world its own safe-deposit box to store their selected seeds regardless of epidemics, terror, war and other earthly catastrophes from Hawking's list. Four million seeds mainly of edible plants were stockpiled in this stony archive, a frozen Eden, to secure worldwide agricultural and food supplies isolated from the fate of the earthly environment.[35]

In 2000 the American novelist T. C. Boyle envisioned a future in which 'the environment is all indoors now anyway, right on down to the domed fields that produce the arugula for [our] salads and the four-walled space [we call] home'.[36] *A Friend of the Earth* portrays a near-future world that is the antithesis to an ecologically well-balanced site like the seed ark. Uncontrollable extreme weather conditions, storms, floods and droughts reflect the present apprehension about global climate change. The novel describes how within three decades a large part of the earth's landmass has turned uninhabitable and uncultivable. But the author is more interested in the extreme social divides that environmental change entails. The few young and rich control the few truly luxurious comfortable, air-conditioned and well-protected estates. Outside of these refuges rage the storms of anarchy. Social security and public health systems have long been done away with. A growing number of old and poor people have to fend for themselves, while in their bouts of magnanimity and eccentric whims of charity the rich attend to the conservation of the remaining rare animal species on earth.

Established at one time to protect nature from human intrusion, today's nature reserves serve as retreats, as arks and islands in a rising sea of disorder. Globalization critic Naomi Klein has argued that such retreats necessarily emerge with 'disaster capitalism'. She calls them global 'green zones'.[37] Following Klein, green zones articulate and implement a growing global apartheid system separating environmental refugees, economic asylum seekers and the migrating poor from the wealthy and well-shielded elite privileged enough to keep up their privatized and highly secured infrastructure. One such embodiment of Klein's global green zone is aptly called *The World*. This largest privately owned yacht on the planet offers luxury ocean residences to those who wish to 'Travel the world without leaving home'.[38] This ark does not wait for anyone and anything, nor does it go anywhere. Like Nemo on the *Nautilus*, its passengers are cosmopolitans of no nationality. They live in the world at large.

Ships still fare well in today's imagination. Seed arks protect natural capital from the unruly environment; luxury liners protect human capital from the unruly poor. And spaceships set out not to explore a vast and unknown universe but to monitor the earth. Like an extension of *Silent Running*, the movie *WALL•E* in 2008 featured the spaceship *Axiom* that took mankind on an indefinite cruise in space as toxicity levels on earth became too high for human habitation. The lone rusty robot WALL-E (Waste Allocation Load Lifter – Earth Class) tends to a devastated and lifeless earth by diligently compiling and compacting trash. The movie presents an assembly of familiar themes from mythology and science fiction. The Robinson Crusoe motif comes alive with the junk that WALL-E collects and carefully stores, from musical videos and show tunes to Rubik's cube, and the young seedling he stumbles on. The island motif blends with the theme of a possible future paradise when WALL-E meets

EVE (Extraterrestrial Vegetation Evaluator), a feminized probe of advanced technology sent out from space to search for life on earth and to explore human-kind's possibilities of repopulating the earth. Director Andrew Stanton himself pointed to the spaceship *Axiom* taking the role of a modern-day Noah's ark. In her versatile way EVE acts not only as first woman but also as the biblical white dove; she incorporates the plant and immediately adopts an egg-like form, incubating the idea that 'Earth has been restored to a life-sustaining status'.[39]

The movie's plot revolves around a scheme maintained by the main board computer named Autopilot and his directive to cancel 'Operation Recolonize' and to keep humanity in space. Autopilot argues that 'on the *Axiom* you will survive', to which the Captain replies: 'I don't want to survive. I want to live!' In this and other expressions of a back-to-earth movement, Spaceship Earth was almost forgotten. No longer motivating political and popular culture it continued to exist only in the nooks of the biosphere and space sciences. Only recently Spaceship Earth had a hesitant comeback in anthropogenic climate change research and in the discourse of the 'Anthropocene' as the new geological age character-ized by humanity's force.[40] A closer look into a combination of circumstances located around 1990 is needed in order to understand why the spaceship was abandoned. Around 1990 is the time when coherent and encompassing uto-pias of time and space lost their meaning and the views of what to expect in the future, what would be fleeting and what would last, what should be preserved and what would be dispensable thoroughly shifted. The historical framework was set with the fall of the Berlin Wall in 1989. As former national territories in the East broke up and territorial boundaries were rearranged and renewed, the notion of globality that had defined the Cold War situation of a threatened but ideologically stable environment lost its capacity for holding the world together around a core of affiliations. After 1990 a common worldview that took hold was global neoliberal capitalism, ideologically embedded in the victorious dem-ocratic regimes.[41]

This condition has been barely disguised by the term 'globalization' that emerged by emphasizing global financial markets in the 1980s, and which entered public consciousness in the early 1990s. A neoconservative and also decidedly anti-environmental political agenda was enforced in the Western world under the Reagan and Thatcher administrations in the 1980s. The discourse of glo-balization optimistically accented global connections and interdependencies by referring to the intricate networks of communication, transport, financial transactions and the promise of gains for all within the New Economy. But glo-balization also became another word for imposed homogeneity and ubiquity. What has been called the 'end of history' in the scholarship on the postmod-ern condition, a phrase which referred to the end of the grand universalizing narratives of a common heritages and futures, was complemented by the end of geography, that is to say the collapse of the universal master maps.[42] As the globe grew increasingly pluralized and individualized, a modern myth like Spaceship Earth could have only nostalgic character.

Post-Normal Science

The humanities likewise attended to the newly discovered webs of local traditions and values, questioning the universal scope and reach of scientific truths. As the collective and contingent character of facts became an object of research, the longstanding notion of scientific problem-solving activity in a sectorally organized academic establishment – 'normal science' in the Kuhnian sense – lost its ground and scope. Around 1990, the temporal as well as the geographic extent of scientific facts was reduced and replaced by an approach toward research and policy that acknowledged locality, uncertainty and complexity. In the early 1990s, methodologist Silvio Funtowicz and philosopher of science Jerome Ravetz termed the explorative, cooperative and intrinsically transdisciplinary scientific activity 'post-normal science'.[43] Michael Gibbons and colleagues from social science studies introduced a similar methodology in the mid-1990s that they called Mode 2 research, which also highlighted the problem-based, context-laden and multidisciplinary ways of knowledge production.[44] The 'Science Wars', the outright aggressive trench warfare between natural and social scientists triggered in 1996, was only one very explicit articulation of a new situation in which hard scientific facts softened while soft political values hardened.[45] Facing a quake of the disciplinary foundations that for a long time had sustained the scientific, social and moral order and authority, natural scientists set out to prove that social scientists and humanities scholars had no say in the matter of nature.

Struggles about the sovereignty over the world and the earth also affected the environmental sciences. Constructivist tendencies seemed to deprive pro-environmental political actions of their grounds and to play into the hands of blatant climate change deniers. The environmental destruction indulged in by humanity, so the accusation, was topped by the environmental deconstruction indulged in by the humanities.[46] Bruno Latour employed this provocation to point out that it will not do to squander the idea of exact knowledge and force the sciences to give up on their monopoly on nature without a sound plan to assess and revaluate the societal 'matters of concern' that used to always come second. Wolfgang Sachs as well proposed that to 'count', to derive knowledge from the quantitative examination of nature, should no longer be of greater importance than to 'account', to argue, justify and contextualize research results.[47] Both claims urged academia not to hunt for a new metaphysics but rather to develop a new 'infra-physics'.[48] Latour's term of infra-physics insinuates that safety will not be recovered by developing new ontologies to replace the metaphysical foundations of modernity, but by creating new infrastructures to allow the circulation and handling of the many new questions and objects located between the networks of science and of politics, between matter and meaning, and between nature and culture.

To the degree that modern societies were geared towards stability, order, maintenance and control, their elaborate infrastructures enlisted environments in unprecedented ways and scales and exposed intricate technological architectures as fragile and vulnerable. Oil spills, melting icecaps, hurricanes and unprecedented epidemics: Latour calls the new environmental phenomena 'collective experiments', and Wolfgang Krohn speaks of 'reality experiments'.[49] The environments involved in these large-scale real-time experiments are dynamic and powerful, enhanced, infectious and toxic, and they are highly politicized. They may be idealized, but they are not inert or immutable or mute objects of negotiation. They call for caution, but they are hardly suited for moral appeals. In whatever way these issues will be resolved, it is unlikely that humanity will again board Spaceship Earth and similar vehicles transporting modern visions of a rational environment and technoscientific mastery.

Sustainable Growth

In 1986 the German sociologist Ulrich Beck put forth his case of modern societies being increasingly confronted with the effects of their sophisticated technologies. Beck maintained that modernity and uncertainty stand in a paradoxical, even mutually constitutive relationship. In the way that modern societies have sought to eliminate uncertainties, minimize threats, reduce errors and accidents and contain exceptions, the disturbances have nevertheless been proliferating. The 'risk society' was Beck's term to describe modern societies and the ways in which they generate 'risks' from dangers and hazards. Compared to providence or destiny, risk is a very modern concept. Based on probabilistic theory, risk rationalizes and operationalizes chance and uncertainty by extracting regular patterns from disorder and disaster. As an object of statistical quantification and calculation, risk can be assessed, managed, insured and capitalized. Yet Beck's risk society neither entailed nor expected nor did it allow for a clear scheme of safety control: risk is open, permits different outcomes, and it embraces the normal as well as the exceptional, which is a crucial condition to Beck's argument that risks no longer affect a community in the form of external threats from the outside, but emerge from within: the notion of 'reflexivity' entered the debate on modernity. Beck expanded his concept to the 'world risk society' to stress that local actions are not locally bound in their consequences anymore but have ubiquitous effects.[50]

In 1987 a vision of a participatory, bottom-up form of social-environmental governance materialized in the program of Sustainable Development. The prominent report *Our Common Future* by the World Commission on Environment and Development headed by the Norwegian Prime Minister Gro Harlem Brundtland coined the celebrated definition of 'sustainable development' as a 'development that meets the needs of the present without compromising the ability of future generations to meet their own needs'.[51] Next to 'needs' as a key concept, the definition also touched on 'limitations'. Notably, the study did not refer to environmental limits as earlier reports like *The Limits to Growth* had, but to the 'limitations imposed by the state of technology and social organization on the environment's ability to meet present and future needs'.[52] The Brundtland Report did not waste too much space on environmental doom-saying. Instead the report suggested seeing limits not as absolute but as relative, not as defined by nature but as defined and governed by society. Limits conveyed the challenge of proficient political organization and efficient application of scientific-technological means. Accordingly, the report promoted the belief that the earth's carrying capacity could in principle be raised by further technological progress.

Also unlike the 1972 report to the Club of Rome, the Brundtland report emphasized locality and diversity over ideas of a uniform planetary development and management, and it accepted the politics of small steps over a master plan. It called for a politics of the earth and a notion of 'stewardship' that abandoned the grand modernist ideas of governing the future, even though the rhetoric outlining the problems was familiar. With a 'call for action' the commission sought to merge the perceived risks and crises of environment and development into one single common problem: 'The Earth is one but the world is not'.[53] The assessment implied a physically unified and well-studied planet and a world conceived of as diverse. Making ample use of performative statements of unity, the report attempted to map the politically disparate world onto the physically integrated planet, without abstaining from stressing 'common' concerns, challenges and endeavours in a normative way. It set the primary direction as being 'From One Earth to One World'.[54]

Risk analysis, risk evaluation and risk management were seen as crucial strategies to come to grips with the unintended environmental and social consequences of human action. Environmental risks on the global scale were growing faster than the social capacities to handle them. Experts from all fields of science and society and from all areas of the world were called upon to raise and collect data to gain a global picture. The commission was concerned less with the physical environmental problems than with the inner conflicts and contradictions of an international community of formally equal organizations that acted in favour of a common heritage but were entangled in competitions about global leadership. The call for consultative and collaborative global capacity building processes is an example of the shift from governmental regimes powered by strong nation-states towards forms of global governance understood as broad configurations of state organizations and civil society that incorporated business and professional interests and included public-private partnerships.

In a striking difference to the future studies of the 1970s, *Our Common Future* restored growth as a respectable aim of environmental politics. As the limits regarding resources and the capacity of the earth's biosphere to cope with human impacts were conceived as technological and societal in character, and thus provisional or conditional on scientific knowledge and political insight, they had to be subjected to far-sighted innovation policies. The study particularly listed 'reviving growth' as a 'strategic imperative' to develop the world as a whole in a sustainable way.[55] 'Sustainable growth' has since set a new moral imperative for the world economies. The concept exhibits some relation to the ecological sufficiency and efficiency revolutions that economists like Boulding and Fuller had foreseen for the earth in the 1960s. The program of sustainable development as brought forth by the Brundtland Report expressed the idea that economic, scientific and technological progress will entail environmental benefits through increased resource efficiency. Deliberations of ecological efficiency or 'ecological modernization' held that ecological issues need not be pursued against economical issues but as a part of a new era of economic growth.[56]

'Is small still beautiful?' Wolfgang Sachs rhetorically countered with regard to the 1973 book *Small Is Beautiful* in which Ernst Schumacher had proposed the downscaling of the economy and industry to human proportions and sustainable dimensions.[57] The earth at the beginning of the twenty-first century seemed to be characterized to a great extent by the globalization of financial markets and a short-lived consumer culture. In the 'age of unrestricted mega-economy' of cash flows and commodity flows, the only elements that seemed sustainable and that accordingly could be exchanged in standardized ways were the units of global trade and global transport: capital and containers. It was only appropriate that in 2005 an international edition of the magazine *Der Spiegel* on 'Globalization: The New World' recaptured the famous NASA picture of Earthrise in the form of a standard shipping container (see Figure 6.1).

The quotation of the earth 'rising' over the surface of the moon as seen by the crew of Apollo 8 in 1968 conveys the familiar motif of the vulnerable Blue Planet floating in space. But the fragility of the earth blends with the terrific possibilities of containerization that have sped up and scaled up the commodity flows around the globe. Invisible in the background are the automated technical processes of handling and shipping cargo profitably. Implicit are the onsite technologies of stocking up and of computerized distribution, allocation and disposal of freight.

Figure 6.1: Shipped Earth. *Spiegel* **Special No. 7/2005, International Edition 'Globalization: The New World', front page.**

Invisible are those who gain and those who lose in the intricate webs and contingent movements of financial capital. The risky and often random enterprises anticipate boom but take crash into account. The picture displays a success story. At the same time it forsakes the notion of an integrated and mastered planet by placing the earth itself at stake. The earth has become an object boxed and handled much like any other perishable consumer good. It is precious only insofar as it can be shipped to any place and sold for the best offer. Global inequalities seem to pale in light of the power of global markets. Yet the imprint repeats the asymmetrical historical and geographical order of 'this side up'.[58]

Population Decline

On Tuesday, 17 October 2006, at 7:46 a.m. Eastern Time, the world witnessed a historical moment: The population of the United States of America reached 300 million people. The Population Clock chiming in the US Census Bureau was a cause for celebration and anxiety.[59] In terms of absolute numbers, the new citizen was highly welcomed. The US takes pride in being the third strongest nation on earth, demographically, directly following China and India. Presently, the US is the only industrial nation whose population is actually growing. Proclaiming a 'net gain' of one person every 13 seconds corresponds to the traditions of a national political economy assessing the power of the state in terms of the numerical strength of its population. Growth is the unquestioned standard against which a country's power is measured. But the faceless child of statistical calculations has also given rise to concerns. When and how this new American actually entered the country is not known. It might have been a newborn, but more likely it was an immigrant stepping off a plane or boat or illegally crossing the border. If it were a baby it might have been of white Anglo-Saxon protestant descent, but more likely it was of one of the many other colours and religions that are on their way of holding the majority in the country.[60]

The American example is just one of the paradoxes of a new situation in which 'underpopulation' in the West meets 'migration pressures' from poorer parts of the world. Today there are twice as many people on earth as there were during the 1960s, and that number is increasing. In many fields, particularly in developmental politics, world population is still considered a problem of global dimensions. The language of the world population growth discourse, however, has lost some of its dramatic tone and political impact.[61] The industrial countries in the Northern Hemisphere currently suffer from new fears of 'shrinkage'. In Europe and Japan, the 'demographic transition', the move from high to low birth rates anticipated for highly industrialized countries, has led to dismal forecasts of a shortage of young and an excess of old people; the bleakest anticipations have warned of the depopulation of entire regions.[62]

> We should have been warned in the early 1990s. As early as 1991 a European Community Report showed a slump in the number of children born in Europe – 8.2 million in 1990, with particular drops in the Roman Catholic countries. We thought that we knew the reasons, that the fall was deliberate, a result of more liberal attitudes to birth control and abortion, the postponement of pregnancy by professional women

pursuing their careers, the wish of families for a higher standard of living. And the fall in population was complicated by the spread of AIDS, particularly in Africa. Some European countries began to pursue a vigorous campaign to encourage the birth of children, but most of us thought the fall was desirable, even necessary. We were polluting the planet with our numbers; if we were breeding less it was to be welcomed.

What reads like another sober analysis of the calamity of depopulation in the West is a paragraph taken from the novel *The Children of Men* published by the British author P. D. James in 1992.[63] The book tells a dystopian story of a world without offspring. Although she wrote the novel shortly before the spectre of population decline fully hit the Western world, James proves to be very perceptive of the many different themes of the contemporary population discourse:

> Most of the concern was less about a falling population than about the wish of nations to maintain their own people, their own culture, their own race, to breed sufficient young to maintain their economic structures.[64]

James, born in the 1930s and a fairly conservative author herself, overstates the feared social effects of depopulation and aging by staging a sudden worldwide fertility crisis. In 1994, practically overnight no more children are born and even the sperm frozen for experiments and artificial inseminations loses its potency. The youngest people on earth, those born in the year 1995, are called the Omegas. In the near future the Omegas will be the last people on earth. The reasons for the abrupt strike of infertility are unclear and remain unresolved. James places the triggering events in the past to confront the readers with the effects on a world that has lost its power to breed for several decades. She sets her story in year 2021. In contrast to the movie *Zardoz* where immortality makes people yield to lethargy, it is inevitable death that makes people succumb to resignation and apathy. The book is concerned with traditional values and saturated with philosophical and religious overtones. Its title, 'The Children of Men', is taken from a funeral psalm of biblical origin (psalm 90:3) that in the book reads, 'Thou turnest man to destruction: again thou sayest, Come again, ye children of men. For a thousand years in thy sight are but as yesterday'.[65]

In 2021 the British population has shrunken by half. In great detail the middle-aged historian and university lecturer Theo conveys 'the small domestic concerns of a shrinking world'[66] and how they affect his and his fellow citizens' everyday life. One focus is on the local forms of governance that develop under the growing global crisis. A Council rules the country in a pseudo-democratic government that exhibits totalitarian traits. Free elections are a matter of the past. The Council keeps up a rigid border control and compulsory recurring fertility testing of every mature person. The authoritarian regime seems nevertheless attractive to many citizens since it promises in a paternalistic way to care for the people until the very end. Effects of population decline have long become visible and perceptible. Schools have been closed, universities teach only the elderly, and roads are no longer repaired. Workers have to be imported to get the basic infrastructural work done. The Sojourners are kept like slaves without contracts and without civil rights. Great Britain profits as it did in the past from global inequalities in economic and technological power.

To suppress meaninglessness and monotony people try to keep up what passes as a daily normal life. They make do with surrogate children, dolls made from porcelain that are available in different ages, or sizes respectively, in a twisted similarity to the robot babies that couples are offered to purchase in the movie *Z.P.G.* (*Zero Population Growth*). In twenty-first century Britain there are pseudo-births, quasi-mothers and christenings. James also explores potential problems that an aging society may pose to a decaying infrastructure. She depicts some drastic measures that remind the reader of *Soylent Green* and the actions taken against overpopulation in the movie; these are illustrated when she writes that 'the Council now pay[s] handsome pensions to the relations of the incapacitated and dependant old who kill themselves'.[67] Not only in an overpopulated but also in a depopulated world it is possible to conceive of old people as a societal burden. The practice of 'Quietus', institutionalized by the government, advises old people to a voluntary death, an appeal that gains additional weight by financially rewarding their relatives. The old people consenting to die set out on their last journey in boats. They are collectively rowed out to the open sea to be drowned. In this story the ship appears as the vehicle of death.

James's book illustrates how the terms of the population discourse have changed since the 1970s. It also shows that the rigid and conservative logic of population accountancy remains in effect, along with its inherent threats and constraints. While current population summits continue to discuss high birth rates in the global South, the North no longer castigates but rigorously demands accumulation and multiplication. The book only hints at the ways in which the emerging discourse of population decline reflects and echoes the terms of population storage and efficiency discussed in chapter 4 of this book, and it only touches on how the population question has been placed into different contexts of national security, global capitalism, global environmental degradation and global economic and environmental migration. The relentless defence of the US border to Mexico and of Fortress Europe shows that migration and immigration are still rejected as possible solutions to inequalities in world populations with regard to the concentration of people and of wealth. Migration is instead looked upon as the escalation of the deterioration of the North. The related security debates resonate with notions and terms of bio-economic performance as former white majorities feel threatened by African, Hispanic, Arab or Asian communities, in fear of the draining of collective and individual capital and of the overthrowing of longstanding social contracts by cohorts of 'foreigners'.[68]

The Lifeboats of Tomorrow

The movie *Children of Men* was released in 2006, when debates on underpopulation peaked.[69] In noteworthy contrast to the novel, however, the movie is interested neither in possible individual nor in the social effects of depopulation. Instead director Alfonso Cuarón uses population issues to concentrate on present effects of globalization at large. Set in the year 2027, eighteen years after the last child was born on earth, resignation and deterioration have spread. Unlike the tale told in the book, no place remains that could remotely pass for normal. Great Britain is in the grip of civil wars and fundamentalist terror. Rage, open violence and random destruction dominate the streets of London; bomb explosions of terrorist acts frequently strike bustling sites. 'The world has collapsed but Britain soldiers on': The government propagates messages of perseverance to its people while sending out armed troops to enforce interior sovereignty and to defend the state borders against illegal immigrants. Britain has become the aspiration of hundred thousands of 'fugees' of war, environmental disaster and economic hardship.

A mixture of pleasure and fear holds the capital in a tight grip. An all-pervasive industry of electronic media entertainment for the wealthy goes hand-in-hand with a massive police surveillance of the poor. Theo's cousin, a minister of the state, lives in London's Tate Modern/Battersea Power Station in the midst of the world's finest art, which has become meaningless but for the few who can afford to surround themselves with the impressive sculptures and paintings of the past. In the streets, filth reigns. Power excavators shovel tons of garbage away from bleak apartment buildings. With the environmental degradation portrayed, the movie departs from James's original. The book pictures how nature regenerates in the midst of social regression, filling up the ample space left by the retreating humans. The movie instead takes care to present the world in steel grey and brown colours as if to bluntly expand infertility metaphorically to all aspects of human and natural life.

Like the novel, the movie's plot revolves around a scheme of a guerrilla group engaging Theo to escort a pregnant woman named Kee out of the country. The plan is to keep Kee from falling into the hands of the government and to see her into the care of a global humanitarian organization called the 'Human Project'. Kee and her child are to be brought to a 'Sanctuary' on the Azores. While James is rather careful in her philosophy about the significance of human communities, the movie brings in the big guns to get its message across. The moment when Theo learns that Kee is expecting a child is made into a key scene, set in a barn among the farm animals. In a postmodern version of the story of the birth of Jesus Christ the scene features all the elements of the original Bethlehem situation and more: the outcasts, the makeshift lodgings, the stable, the pregnancy and a black mother: Kee is an illegal immigrant from Africa, a refugee unwanted in the country. The spirituality of the situation also profits from the fact that the movie gets by without sex. And like in the biblical account we never learn (or care) about the origin of this first child expected in eighteen years. We simply take comfort in the hope brought by the one special child that will redeem a doomed human race. This world has literally been waiting for the 'Second Coming'.

The movie makes every possible effort to depart from the discouraging ending of James's novel. In a dramatic flight from the government's forces, Theo and Kee manage to break into a heavily guarded prison camp for illegal immigrants on the British coast in order to catch the Human Project's ship. The setting allows Cuarón to play out the contemporary situation of a state-run prison in which violence, torture and other transgressions have become the normal state of affairs. He expands the prison into a ghetto, an entire former city in whose ruins the unwanted refugees barely survive. On a bare floor in one of the dilapidated buildings, Kee gives birth to a child. In the final scene Theo rows Kee and her newborn girl out to the open sea in a small rowboat to meet the Human Project's disguised hospital ship. This crucial scene could have been modelled on the movie *Z.P.G.*, in which man, woman and infant, the original family, escape from the decayed city in a small dinghy. Interestingly, however, *Children of Men* again provides a postmodern version of the family. The family is made up of three human beings of different skin colour who are hardly bound by ties of blood but share only ties of dependency, of chosen solidarity and of mutual protection. The movie ends on a hopeful note. The hospital ship by the name of *Tomorrow* picks up Kee and her baby girl.

'What I like is that the solution is the boat', the Slovenian philosopher Slavoj Žižek comments in a documentary feature titled *The Possibility of Hope* that was filmed by Cuarón to accompany the movie.[70] Žižek is one of the cultural critics interviewed to discuss how the movie's themes relate to present world society, and interestingly he focuses on the ship. 'What is the definition of the boat? It's that it doesn't have roots. It's rootless. It floats around. That's the solution.' Žižek recognizes the uncertainties and the precariousness of being suspended in time and space, regarding them as a condition of postmodernity that cannot be rejected but needs to be embraced. 'We must really accept how we are rootless. This is, for me, the meaning of this wonderful metaphor, boat. Boat is the solution, boat in the sense of: you accept rootless, free floating. You cannot rely on anything.'

In this story the ship is a place without a place, a site separated from and yet in the midst of all other places, committed to continuous journeying and exposed to the hazards a nomadic existence entails. There may be similarities to the ark, and to the spaceship, but Žižek's postmodern standpoint involves none of the modern ideas of direction and control – neither a coherent space nor a common view of the world nor security, and least of all arrival:

'You know, it's not a return to land'.

WORKS CITED

'30 Years of ZPG', *Reporter*, December 1998, pp. 12–19, *The Population Connection*, at http://www.populationconnection.org/site/DocServer/1219thirtyyears.pdf?docID=261 [accessed 28 August 2014].

Abbott, C., *Frontiers Past and Future: Science Fiction and the American West* (Lawrence, KS: University Press of Kansas, 2006).

Achterhuis, H. J., 'Van Moeder Aarde tot ruimteschip: humanisme en milieucrisis', in H. J. Achterhuis, *Natuur tussen mythe en techniek* (Baarn: Ambo, 1995), pp. 41–64.

Adams, W. M., *Green Development: Environment and Sustainability in the Third World* (London and New York: Routledge, 1992).

Agamben, G., *Homo Sacer: Sovereign Power and Bare Life* (Stanford, CA: Stanford University Press, 1995).

—, *State of Exception* (Chicago, IL, and London: University of Chicago Press, 2005).

Agrawal, A., *Environmentality: Technologies of Government and the Making of Subjects* (New Delhi: Oxford University Press, 2005).

Akera, A., *Calculating a Natural World: Scientists, Engineers, and Computers During the Rise of U.S. Cold War Research* (Cambridge, MA, and London: MIT Press, 2006).

Allen, G. E., 'Old Wine in New Bottles: From Eugenics to Population Control in the Work of Raymond Pearl', in K. R. Benson, J. Maienschein and R. Rainger (eds), *The Expansion of American Biology* (New Brunswick, NJ, and London: Rutgers University Press, 1991), pp. 231–61.

Allen, J., *Biosphere 2: The Human Experiment* (New York: Penguin Books, 1991).

—, *Me and the Biospheres: A Memoir by the Inventor of Biosphere 2* (Santa Fe, NM: Synergetic Press, 2009).

Allen, J., and M. Nelson, *Space Biospheres* (1986; Oracle, AZ: Synergetic Press, 1989).

—, 'Biospherics and Biosphere 2, Mission One (1991–1993)', in B. D. V. Marino and H. T. Odum (eds), *Biosphere 2: Research Past and Present* (Amsterdam: Elsevier Science, 1999), pp. 15–29.

Allen, J., M. Nelson and A. Alling, 'The Legacy of Biosphere 2 for the Study of Biospherics and Closed Ecological Systems', *Advances in Space Research: The Official Journal of the Committee on Space Research* (COSPAR), 31 (2003), pp. 1629–39.

Allen, J., T. Parrish and M. Nelson, 'The Institute of Ecotechnics: An Institute Devoted to Developing the Discipline of Relating Technosphere to Biosphere', *Environmentalist*, 4 (1984), pp. 205–18.

Alling, A., and M. Nelson, *Life Under Glass: The Inside Story of Biosphere 2* (Oracle, AZ: Space Biospheres Ventures and The Biosphere Press, 1993).

'A New Set of National Priorities: Three Blueprints for Postwar Reconciliation and Reconstruction', *Playboy*, 18 (1971), pp. 146 ff.

Andrews, J. T., and A. A. Siddiqi, *Into the Cosmos: Space Exploration and Soviet Culture* (Pittsburgh, PA: University of Pittsburgh Press, 2011).

Andrews, R. N. L., *Managing the Environment, Managing Ourselves: A History of American Environmental Policy* (New Haven, CT, and London: Yale University Press, 1999).

Anker, P., *Imperial Ecology: Environmental Order in the British Empire, 1895–1945* (Cambridge, MA, and London: Harvard University Press, 2001).

—, 'The Closed World of Ecological Architecture', *Journal of Architecture*, 10 (2005), pp. 527–52.

—, 'The Ecological Colonization of Space', *Environmental History*, 10 (2005), pp. 239–68.

—, 'Buckminster Fuller as Captain of Spaceship Earth', *Minerva*, 45 (2007), pp. 417–34.

—, *From Bauhaus to Ecohouse: A History of Ecological Design* (Baton Rouge, LA: Louisiana State University Press, 2010).

Appadurai, A., *Modernity at Large: Cultural Dimensions of Globalization* (Minneapolis, MN: University of Minnesota Press, 1996).

Bachelard, G., *The Poetics of Space: The Classic Look at How We Experience Intimate Places* (1958; Boston, MA: Beacon Press, 1994).

Baker, P., *The Story of Manned Space Stations: An Introduction* (London et al.: Springer, 2001).

Barbier, E. B., J. C. Burgess and C. Folke, *Paradise Lost? The Ecological Economics of Biodiversity* (London: Earthscan, 1994).

Barnes, J., *A History of the World in 10½ Chapters* (1989; New York: Vintage, 1990).

Barthes, R., 'The *Nautilus* and the Drunken Boat', in R. Barthes, *Mythologies* (1957; New York: Hill and Wang, 1972), pp. 65–7.

—, 'Myth Today', in R. Barthes, *Mythologies* (1957; London: Jonathan Cape, 1972), pp. 109–59.

Bashford, A., *Global Population: History, Geopolitics and Life on Earth* (New York: Columbia University Press, 2014).

Baudrillard, J., 'Biosphère II', in A.-M. Eyssartel and B. Rochette (eds), *Des mondes inventés: les parcs à thème* (Paris: Éditions de la Vilette, 1992), pp. 127–30.

—, *Simulacra and Simulation* (1981; Ann Arbor, MI: University of Michigan Press, 1994).

—, 'Maleficient Ecology', in J. Baudrillard, *The Illusion of the End* (1992; Cambridge: Polity Press, 1994), pp. 78–88.

—, 'Überleben und Unsterblichkeit', in D. Kamper and C. Wulf (eds), *Anthropologie nach dem Tode des Menschen: Vervollkommnung und Unverbesserlichkeit* (Frankfurt: Suhrkamp, 1994), pp. 335–54.

Bauman, Z., *Intimations of Postmodernity* (London and New York: Routledge, 1992).

Bayly, C. A., *The Birth of the Modern World: Global Connections and Comparisons, 1780–1914* (Oxford: Blackwell, 2004).

Beck, U., *Risk Society: Towards a New Modernity* (1986; London: Sage, 1992).

—, *World Risk Society* (1997; Malden, MA: Polity Press, 1999).

Beck, U., A. Giddens and S. Lash, *Reflexive Modernization: Politics, Tradition and Aesthetics in the Modern Social Order* (Stanford, CA: Stanford University Press, 1994).

Beinart, W., and L. Hughes, *Environment and Empire* (Oxford and New York: Oxford University Press, 2007).

Belasco, W., 'Algae Burgers for a Hungry World? The Rise and Fall of Chlorella Cuisine', *Technology and Culture*, 38 (1997), pp. 608–34.

Bentham, J., *The Works of Jeremy Bentham* (Edinburgh: William Tait, 1838–43).

Bergermann, U., I. Otto and G. Schabacher (eds), *Das Planetarische: Kultur – Technik – Me-*

dien im postglobalen Zeitalter (München: Wilhelm Fink, 2010).

Bernal, J. D., *The World, The Flesh and The Devil: An Inquiry into the Future of the Three Enemies of the Rational Soul* (1929; London: Jonathan Cape, 1970).

—, *The Origin of Life* (London: Weidenfeld & Nicolson, 1967).

'Beyond Earth's Boundaries: Human Exploration of the Solar System in the 21st Century.' 1988 Annual Report to the Administrator, National Aeronautics and Space Administration, Washington, DC, Office of Exploration 1988, *Educational Resources Information Center (ERIC)*, at http://eric.ed.gov/ERICDocs/data/ericdocs2sql/content_storage_01/0000019b/80/1f/74/52.pdf [accessed 28 August 2014].

'Big Science at Biosphere 2' (2014), *University of Arizona*, at http://www.arizona.edu/big-science-biosphere-2 [accessed 28 August 2014].

Bilstein, R. E., *Orders of Magnitude: A History of NACA and NASA, 1915–1990* (Washington, DC: National Aeronautics and Space Administration, Scientific and Technical Information Division, 1989).

Bloomfield, B. P., *Modelling the World: The Social Constructions of Systems Analysts* (Oxford: Basil Blackwell, 1986).

Blumenberg, H., *Arbeit am Mythos* (Frankfurt: Suhrkamp, 1979).

—, *Schiffbruch mit Zuschauer: Paradigma einer Daseinsmetapher* (1979; Frankfurt: Suhrkamp, 1997).

—, 'Das Jahr 1969: Mondbezwingung und Umweltschutz', in H. Blumenberg, *Die Vollzähligkeit der Sterne*, (Frankfurt: Suhrkamp, 2000), pp. 439–40.

Borgstrom, G., *Too Many: A Study of Earth's Biological Limitations* (London: Macmillan, 1969).

Boulding, K. E., *The Meaning of the Twentieth Century: The Great Transition* (1964; New York: Harper & Row, 1965).

—, 'The Economics of the Coming Spaceship Earth', in H. Jarrett (ed.), *Environmental Quality in a Growing Economy*, Essays from the Sixth RFF Forum on Environmental Quality held in Washington 8 and 9 March 1966 (Baltimore, MD: Johns Hopkins Press, 1966), pp. 3–14.

—, *Beyond Economics: Essays on Society, Religion, and Ethics* (1968; Ann Arbor, MI: University of Michigan Press, 1970), pp. 275–87.

—, 'Spaceship Earth Revisited', in H. E. Daly (ed.), *Economics, Ecology, Ethics: Essays Toward a Steady-State Economy* (San Francisco, CA: W. H. Freeman and Company, 1980), pp. 264–6.

Bourguet, M.-N., C. Licoppe and H. O. Sibum, *Instruments, Travel and Science: Itineraries of Precision from the Seventeenth to the Twentieth Century* (London: Routledge, 2002).

Bowker, G., 'How to be Universal: Some Cybernetic Strategies, 1943–1970', *Social Studies of Science*, 23 (1993), pp. 107–27.

Boyle, T. C., *A Friend of the Earth* (2000; London: Bloomsbury, 2001).

Brant, S., *The Ship of Fools* (1494; New York: Dover, 1962).

Briden, J. C., and T. E. Downing (eds), *Managing the Earth: The Linacre Lectures 2001* (Oxford and New York: Oxford University Press, 2002).

Brown, L. R., *The Twenty-Ninth Day: Accommodating Human Needs and Numbers to the Earth's Resources* (A Worldwatch Institute Book) (New York: W. W. Norton, 1978).

Brunhes, J., *Les Limites de notre cage*, Discours à l'occasion de l'inauguration solennelle des cours universitaires, le 15 novembre 1909 (Fribourg, 1911).

Buckminster Fuller, R., 'The Case for a Domed City', *St. Louis Post Dispatch* (26 September 1965), pp. 39–41.

—, *Operating Manual for Spaceship Earth* (1969; New York: E. P. Dutton & Co., 1971).

—, 'Noah's Ark #2', in J. Krausse and C. Lichtenstein (eds), *Your Private Sky: Richard Buck-minster Fuller, The Art of Design Science* (Baden: Lars Müller, 1999), pp. 176–226.

Buell, F., *From Apocalypse to Way of Life: Environmental Crisis in the American Century* (New York and London: Routledge, 2003).

Bush, V., *Science – The Endless Frontier* (Washington, DC: United States Government Printing Office, 1945).

Butler, J., *Bodies that Matter: On the Discursive Limits of 'Sex'* (New York and London: Routledge, 1993).

Butrica, A. J., *The Navigators: A History of NASA's Deep-Space Navigation* (CreateSpace Publishing, 2014).

Callicott, J. B., and M. P. Nelson (eds), *The Great New Wilderness Debate: An Expansive Collection of Writings Defining Wilderness from John Muir to Gary Snyder* (Athens, GA: University of Georgia Press, 1998).

Cannon, S. F., *Science in Culture: The Early Victorian Period* (New York: Dawson, 1978).

Carr-Saunders, A. M., *The Population Problem: A Study in Human Evolution* (1922; New York: Arno, 1974).

Carson, R., *Silent Spring* (Boston, MA: Mifflin, 1962).

Catton, W. R., *Overshoot: The Ecological Basis of Revolutionary Change* (Urbana, Chicago, IL, and London: University of Illinois Press, 1980).

Chakrabarty, D., *Provincializing Europe: Postcolonial Thought and Historical Difference* (Princeton, NJ: Princeton University Press, 2000).

Chamberlain, W. M., 'Population Control: The Legal Approach to a Biological Imperative', *California Law Review*, 58 (1970), pp. 1414–43.

Chaney, S., *Nature of the Miracle Years: Conservation in WestGermany, 1945–1975* (New York and Oxford: Berghahn Books, 2008).

Clynes, M. E., and N. S. Kline, 'Cyborgs and Space', *Astronautics* (September 1960), pp. 26–7, 74–6.

Coates, P., *Nature: Western Attitudes since Ancient Times* (Cambridge: Polity Press, 1998).

Cole, H. S. D., C. Freeman, M. Jahoda and K. L. R. Pavitt (eds), *Models of Doom: A Critique of The Limits to Growth, with a Reply by the Authors of The Limits to Growth* (New York: Universe Books, 1973).

Commoner, B., *Science & Survival* (1966; New York: Ballantine, 1970).

—, *The Closing Circle: Nature, Man & Technology* (1971; New York: Bantam Books, 1972).

Compton, W. D., and C. D. Benson, *Living and Working in Space: A History of Skylab* (Washington, DC: National Aeronautics and Space Administration, 1983).

Connelly, Matthew, 'Population Control is History: New Perspectives on the International Campaign to Limit Population Growth', *Comparative Studies in Society and History*, 45 (2003), pp. 122–47.

—, 'To Inherit the Earth. Imagining World Population, from the Yellow Peril to the Population Bomb', *Journal of Global History*, 1 (2006), pp. 299–319.

—, *Fatal Misconception: The Struggle to Control World Population* (Cambridge, MA, and London: Belknap Press of Harvard University Press, 2008).

Conrad, S., and S. Randeria (eds), *Jenseits des Eurozentrismus: Postkoloniale Perspektiven in den Geschichts- und Kulturwissenschaften* (Frankfurt: Campus, 2002).

Cooper, M., *Life as Surplus: Biotechnology and Capitalism in the Neoliberal Era* (Seattle, WA: University of Washington Press, 2008).

Cosgrove, D., 'Environmental Thought and Action: Pre-modern and Post-modern', *Transac-

tions of the Institute of British Geographers, new series, 15 (1990), pp. 344–58.

—, 'Contested Global Visions: *One-World, Whole-Earth*, and the Apollo Space Photographs', *Annals of the Association of American Geographers*, 84 (1994), pp. 270–94.

—, *Apollo's Eye: A Cartographic Genealogy of the Earth in the Western Imagination* (2001; Baltimore, MD, and London: Johns Hopkins University Press, 2003).

Costanza, R. et al., 'The Value of the World's Ecosystem Services and Natural Capital', *Nature*, 387 (1997), pp. 253–60.

—, 'Changes in the Global Value of Ecosystem Services', *Global Evironmental Change*, 26 (2014), pp. 152–8.

Craige, B. J., *Eugene Odum: Ecosystem Ecologist and Environmentalist* (2001; Athens, GA: University of Georgia Press, 2002).

Crang, M., and N. Thrift (eds), *Thinking Space* (New York and London: Routledge, 2000).

Creighton, M. S., and L. Norling (eds), *Iron Men, Wooden Women: Gender and Seafaring in the Atlantic World, 1700–1920* (Baltimore, MD, and London: Johns Hopkins University Press, 1996).

Creighton, M. S., 'Davy Jones's Locker Room: Gender and the American Whaleman, 1830–1870', in M. S. Creighton and L. Norling (eds), *Iron Men, Wooden Women: Gender and Seafaring in the Atlantic World, 1700–1920* (Baltimore, MD, and London: Johns Hopkins University Press, 1996), pp. 118–37.

Cronon, W. (ed.), *Uncommon Ground: Rethinking the Human Place in Nature* (1995; New York and London: W. W. Norton, 1996).

Cronon, W., 'The Trouble with Wilderness or, Getting Back to the Wrong Nature', *Environmental History*, 1 (1996), pp. 7–28.

Cronon, W., G. Miles and J. Gitlin (eds), *Under an Open Sky: Rethinking America's Western Past* (New York and London: W. W. Norton, 1992).

Cronon, W., G. Miles and J. Gitlin, 'Becoming West: Toward a New Meaning for Western History', in Cronon, William, Miles, George, Gitlin, Jay (eds), *Under an Open Sky: Rethinking America's Western Past* (New York and London: W. W. Norton, 1992), pp. 3–27.

Crosby, A. W., *Ecological Imperialism: The Biological Expansion of Europe, 900–1900* (New York: Cambridge University Press, 1986).

Crutzen, P. J., and E. F. Stoermer, 'The "Anthropocene"', *Global Change Newsletter*, 41 (2000), pp. 17–18.

Crutzen P. et al., *Das Raumschiff Erde hat keinen Notausgang* (Berlin: Suhrkamp, 2011).

Daily, G. C. (ed.), *Nature's Services: Societal Dependence on Natural Ecosystems* (Washington, DC: Island Press, 1997).

Daily, G. C., and P. R. Ehrlich, 'Population, Sustainability, and Earth's Carrying Capacity', *BioScience*, 42 (1992), pp. 761–71.

Daily, G. C., and K. Ellison, *The New Economy of Nature: The Quest to Make Conservation Profitable* (Washington, DC: Island Press and Shearwater Books, 2002).

Dalton, R., 'Biosphere 2 Finds a Buyer', *Nature*, 447 (2007), p. 759.

Daly, H. E. (ed.), *Economics, Ecology, Ethics: Essays Toward a Steady-State Economy* (San Francisco, CA: W. H. Freeman and Company, 1980), pp. 253–63.

Daniels, R. V., *The Fourth Revolution: Transformations in American Society from the Sixties to the Present* (London: Taylor & Francis, 2006).

Dasmann, R. F., *Planet in Peril? Man and the Biosphere Today* (Harmondsworth: Penguin Books – Unesco, 1972).

—, 'Toward a Biosphere Consciousness', in D. Worster (ed.), *The Ends of the Earth: Per-*

spectives on Modern Environmental History (Cambridge: Cambridge University Press, 1988), pp. 277–88.

Daston, L., 'The Moral Economy of Science', in A. Thackray (ed.), *Constructing Knowledge in the History of Science* (*Osiris*, 10 (1995)), pp. 3–24.

Davis, K., 'Population Policy: Will Current Programs Succeed?', *Science*, 158 (1967), pp. 730–9.

—, 'Zero Population Growth: The Goal and the Means', in M. Olson and H. H. Landsberg (eds), *The No-Growth Society* (1973; London: Woburn, 1975), pp. 15–30.

'Definition of Life Support Systems in the Context of the EOLSS', *Encyclopedia of Life Support Systems* (*EOLSS*), developed under the Auspices of the UNESCO, at http://www.eolss.net [accessed 28 August 2014].

Defoe, D., *Robinson Crusoe* (1719; New York: Barnes & Noble Classics, 2005).

DeGroot, G. J., *Dark Side of the Moon: The Magnificent Madness of the American Lunar Quest* (New York: New York University Press, 2006).

Desrosières, A., *The Politics of Large Numbers: A History of Statistical Reasoning*, Cambridge, MA: Harvard University Press, 1998).

Dettelbach, M., 'Humboldtian Science', in N. Jardine, J. A. Secord and E. C. Spary (eds), *Cultures of Natural History* (Cambridge: Cambridge University Press, 1996), pp. 287–304.

DeVorkin, D. H., *Science with a Vengeance: How the Military Created the US Space Sciences after World War II* (New York and Berlin: Springer, 1992).

Dick, S. J., and R. D. Launius (eds), *Critical Issues in the History of Spaceflight* (Washington, DC: NASA History Division, Office of External Relations, 2006).

Die Kolonisierung des Weltraums, by the editors of *Time-Life Books* (Amsterdam: Time-Life Books, 1991).

Dryzek, J. S., *The Politics of the Earth: Environmental Discourses* (Oxford: Oxford University Press, 1997).

Duden, B., 'Population', in W. Sachs (ed.), *The Development Dictionary: A Guide to Knowledge as Power* (London: Zed Books, 1992), pp. 146–57.

Dunlap, T. R., *Faith in Nature: Environmentalism as Religious Quest* (Seattle, WA, and London: University of Washington Press, 2004).

Earth Sciences in the Cold War (Special Guest-Edited Issue, *Social Studies of Science*, 33 (2003)), pp. 629–819.

Edney, M. H., *Mapping an Empire: The Geographical Construction of British India, 1765–1843* (Chicago, IL: University of Chicago Press, 1997).

Edwards, P. N., 'Global Climate Science, Uncertainty and Politics: Data-laden Models, Model-filtered Data', *Science as Culture*, 8 (1999), pp. 437–72.

—, 'The World in a Machine: Origins and Impacts of Early Computerized Global Systems Models', in A. C. Hughes and T. P. Hughes (eds), *Systems, Experts, and Computers: The Systems Approach in Management and Engineering, World War II and After* (Cambridge, MA: MIT Press, 2000), pp. 221–53.

—, *A Vast Machine: Computer Models, Climate Data, and the Politics of Global Warming* (Boston, MA: MIT Press, 2010).

Egan, M., *Barry Commoner and the Science of Survival: The Remaking of American Environmentalism* (Cambridge, MA: MIT Press, 2007).

Ehmer, J., J. Ehrhardt and M. Kohli (eds), *Fertility in the History of the 20th Century: Trends, Theories, Policies, Discourses* (*Historical Social Research/Historische Sozialforschung*, 36 (2011)).

Ehrlich, P. R., *Die Bevölkerungsbombe* (1971; Frankfurt: Fischer, 1973).

—, *The Population Bomb* (1968; New York: Ballantine, 1969).

Ehrlich, P. R., and A. H. Ehrlich, *Population, Resources, Environment: Issues in Human Ecology* (San Francisco, CA: W. H. Freeman, 1970 and 1972).

—, 'Population Control and Genocide', in J. P. Holdren and P. R. Ehrlich (eds), *Global Ecology: Readings Toward a Rational Strategy for Man* (New York: Harcourt Brace Jovanovich, 1971), pp. 157–9.

Ehrlich, P. R., A. H. Ehrlich and J. P. Holdren, *Ecoscience: Population, Resources, Environment* (1970; San Francisco, CA: W. H. Freeman, 1977).

Ehrlich, P. R., and R. L. Harriman, *How to Be a Survivor: A Plan to Save Spaceship Earth* (New York: Ballantine, 1971).

Ehrlich, P. R., and J. P. Holdren, 'Impact of Population Growth', *Science* (1971), pp. 1212–17.

Eigen, M., *Stufen zum Leben: Die frühe Evolution im Visier der Molekularbiologie* (München and Zürich: Piper, 1987).

Eisler, W., *The Furthest Shore: Images of Terra Australis from the Middle Ages to Captain Cook* (Cambridge: Cambridge University Press, 1995).

Elichirigoity, F., *Planet Management: Limits to Growth, Computer Simulation, and the Emergence of Global Spaces* (Evanston, IL: Northwestern University Press, 1999).

Encyclopedia of World Environmental History, ed. S. Krech III, J. R. McNeill and C. Merchant (New York and London: Routledge, 2004).

Enke, S., 'Birth Control for Economic Development', in J. P. Holdren and P. R. Ehrlich (eds), *Global Ecology: Readings Toward a Rational Strategy for Man* (New York: Harcourt Brace Jovanovich, 1971), pp. 193–200.

Erickson, P., J. L. Klein, L. Daston, R. Lemov, T. Sturm and M. D. Gordin, *How Reason Almost Lost Its Mind: The Strange Career of Cold War Rationality* (Chicago, IL: University of Chicago Press, 2013).

Escobar, A., *Encountering Development: The Making and Unmaking of the Third World* (Princeton, NJ: Princeton University Press, 1995).

Etemad, B., *Possessing the World: Taking the Measurements of Colonisation from the 18th to the 20th Century* (New York and Oxford: Berghahn Books, 2007).

Fairchild, H. P., 'Optimum Population', in M. Sanger (ed.), *Proceedings of the World Population Conference, held at the Salle Centrale, Geneva, August 29th to September 3rd, 1927* (London: Edward Arnold & Co., 1927), pp. 72–85.

Fallaci, O., *If the Sun Dies* (1965; New York: Atheneum, 1966).

Fan, F., *British Naturalists in Qing China: Science, Empire and Cultural Encounter* (Cambridge, MA, and London: Harvard University Press, 2004).

Farish, M., *The Contours of America's Cold War* (Minneapolis, MN, and London: University of Minnesota Press, 2010).

Feynman, R. P., 'There's Plenty of Room at the Bottom. An Invitation to Enter a New Field of Physics', *Engineering and Science*, 23 (1960), pp. 22–36.

Fink, C., P. Gassert, D. Junker and D. S. Mattern, *1968: The World Transformed* (Cambridge: Cambridge University Press, 1999).

Flammarion, C., *L'atmosphère: Météorologie populaire* (Paris, 1888).

Fleming, J. R., *Fixing the Sky: The Checkered History of Weather and Climate Control* (New York: Columbia University Press, 2010).

Folsome, C. E., and J. A. Hanson, 'The Emergence of Materially Closed System Ecology', in N. Polunin (ed.), *Ecosystem Theory and Application* (New York: John Wiley & Sons Ltd., 1986), pp. 269–88.

Forrester, J. W., *World Dynamics* (Cambridge, MA: Wright-Allen Press, 1971).

Foucault, M., 'Of Other Spaces', *Diacritics*, 16 (1986), pp. 22–7.

—, *The History of Sexuality*, Vol. 1: *An Introduction* (1976; New York: Pantheon Books, 1978).

—, *Discipline and Punish: The Birth of the Prison* (1975; New York: Vintage Books, 1995).

—, *The Birth of Biopolitics: Lectures at the College de France, 1978–1979* (Basingstoke: Palgrave Macmillan, 2008).

—, *Security, Territory, Population: Lectures at the College de France 1977–1978* (New York: Picador and Palgrave Macmillan, 2009).

Foust, J., 'Google's Moonshot', The Space Review (17 September 2007), at http://www.thespacereview.com/article/957/1 [accessed 28 August 2014].

Fremlin, J. H., 'How Many People Can the World Support?' *New Scientist*, 415 (29 October 1964), pp. 285–7.

Friedman, Y., *Utopies réalisables*, new edn (1974; Paris: Éditions de l'éclat, 2000).

'From 200 to 300 million, how we've changed', *NBC News* (17 October 2006), at http://www.msnbc.msn.com/id/15291977 [accessed 28 August 2014].

Fukuyama, F., *The End of History and the Last Man* (New York and Toronto: Maxwell Macmillan International, 1992).

Funtowicz, S. O., and J. R. Ravetz, 'Science for the Post-Normal Age', *Futures*, 25 (1993), pp. 739–55.

Gabel, M., *Energy, Earth and Everyone: Energy Strategies for Spaceship Earth* (1974; New York: Doubleday, 1980).

—, 'Buckminster Fuller and the Game of the World', incl. 'The World Game – How to Make the World Work' from *Utopia or Oblivion*, in *Buckminster Fuller: Anthology for the New Millennium*, ed. T. T. K. Zung (New York: St Martin's Press, 2001), pp. 122–7, 128–131.

Galison, P., *Big Science: The Growth of Large Scale Research* (Stanford, CA: Stanford University Press, 1992).

—, 'The Ontology of the Enemy: Norbert Wiener and the Cybernetic Vision', *Critical Inquiry*, 21 (1994), pp. 228–66.

Gentry, L., and K. Liptak, *The Glass Ark: The Story of Biosphere 2* (New York: Puffin Books (Viking Penguin), 1991).

Gerhard, U., J. Link and E. Schulte-Holtey (eds), *Infografiken, Medien, Normalisierung: Zur Kartographie politisch-sozialer Landschaften* (Heidelberg: Synchron, 2001), pp. 77–92.

Gerovitch, S., *From Newspeak to Cyberspeak: A History of Soviet Cybernetics* (Cambridge, MA, and London: MIT Press, 2004).

Geyer, M. H., and J. Paulmann (eds), *The Mechanics of Internationalism. Culture, Society and Politics from the 1840s to the First World War* (Oxford: Oxford University Press, 2001).

Gibbons, M., C. Limoges, H. Nowotny, S. Schwartzman, P. Scott and M. Trown, *The New Production of Knowledge: The Dynamics of Science and Research in Contemporary Societies* (London: Sage, 1994).

Gissibl, B., S. Höhler and P. Kupper (eds), *Civilizing Nature: National Parks in Global Historical Perspective* (New York and Oxford: Berghahn Books, 2012).

Gitlin, J., 'On the Boundaries of Empire', in W. Cronon, G. Miles and J. Gitlin (eds), *Under an Open Sky: Rethinking America's Western Past* (New York and London: W. W. Norton, 1992), pp. 71–89.

Goetzmann, W. H., and W. N. Goetzmann, *The West of the Imagination* (New York and London: W. W. Norton, 1986).

Goldsmith, E., R. Allen, M. Allaby, J. Davoll and S. Lawrence, *A Blueprint for Survival* (London: Tom Stacey, 1972).

—, *Blueprint for Survival* (Boston, MA: Houghton Mifflin, 1972).

Golley, F. B., *A History of the Ecosytem Concept in Ecology: More than the Sum of the Parts* (New Haven, CT, and London: Yale University Press, 1993).

Goodwin, C., 'Seeing in Depth', *Social Studies of Science,* 25 (1995), pp. 237–74.

Grab, Y., 'The Use and Misuse of the Whole Earth Image', *Whole Earth Review,* 44 (1985), pp. 18–25.

Gregory, D., *Power, Knowledge and Geography: An Introduction to Geographic Thought and Practice* (Oxford: Blackwell, 2006).

Greiner-Kemptner, U., and R. F. Riesinger (eds), *Neue Mythographien: Gegenwartsmythen in der interdisziplinären Debatte* (Wien, Köln, Weimar: Böhlau, 1995).

Grinevald, J., 'Sketch for a History of the Idea of the Biosphere', in P. Bunyard (ed.), *Gaia in Action: Science of the Living Earth* (Edinburgh: Floris, 1996), pp. 34–53.

Gross, P. R., and N. Levitt, *Higher Superstition: The Academic Left and its Quarrels with Science* (Baltimore, MD: Johns Hopkins University Press, 1998).

Groß, M., H. Hoffmann-Riem and W. Krohn, *Realexperimente: Ökologische Gestaltungsprozesse in der Wissensgesellschaft* (Bielefeld: Transcript, 2005).

Gugerli, D., and B. Orland (eds), *Ganz normale Bilder: Historische Beiträge zur visuellen Herstellung von Selbstverständlichkeit* (Zürich: Chronos, 2002).

Guha, R., *Environmentalism: A Global History* (New York: Longman, 2000).

Guide to Gargoyles and Other Grotesques, ed. A.-C. Fallen (Washington, DC: Washington National Cathedral, 2003).

Hacking, I., *The Taming of Chance* (1990; Cambridge: Cambridge University Press, 2004).

Haeuplik-Meusburger, S., C. Paterson, D. Schubert and P. Zabel, 'Greenhouses and their Humanizing Synergies', *Acta Astronautica,* 96 (2014), pp. 138–50.

Hagen, J. B., *An Entangled Bank: The Origins of Ecosystem Ecology* (New Brunswick, NJ: Rutgers University Press, 1992).

Hagner, M., and E. Hörl (eds), *Die Transformation des Humanen: Beiträge zur Kulturgeschichte der Kybernetik* (Frankfurt: Suhrkamp, 2008).

Hajer, M. A., *The Politics of Environmental Discourse: Ecological Modernization and the Policy Process* (Oxford: Oxford University Press, 1995).

Haller, L., S. Höhler and A. Westermann (eds), *Rechnen mit der Natur: Ökonomische Kalküle um Ressourcen* (*Beiträge zur Wissenschaftsgeschichte,* 37 (2014)).

Hamblin, J. D., *Oceanographers and the Cold War: Disciples of Marine Science* (Seattle, WA: University of Washington Press, 2005).

—, *Arming Mother Nature: The Birth of Catastrophic Environmentalism* (Oxford and New York: Oxford University Press, 2013).

Hanke, J., 'Dive into the new Google Earth', Official Google Blog (2 February 2009), at http://googleblog.blogspot.com/2009/02/dive-into-new-google-earth.html [accessed 28 August 2014].

Haraway, D. J., 'A Cyborg Manifesto: Science, Technology, and Socialist-Feminism in the Late Twentieth Century', in D. J. Haraway, *Simians, Cyborgs, and Women: The Reinvention of Nature* (London: Free Association Books, 1991), pp. 149–81.

—, 'Situated Knowledges: The Science Question in Feminism and the Privilege of Partial Perspective', in D. J. Haraway, *Simians, Cyborgs, and Women: The Reinvention of Nature* (London: Free Association Books, 1991), pp. 183–201.

—, *Modest_Witness@Second_Millenium.FemaleMan©_Meets_OncoMouse™: Feminism and Technoscience* (New York and London: Routledge, 1997).

Hård, M., and A. Jamison, *Hubris and Hybrids: A Cultural History of Technology and Science* (New York and London: Routledge, 2005).

Hardin, G., 'The Tragedy of the Commons', *Science*, 162 (1968), pp. 1243–8.

—, 'Editorial: Parenthood: Right or Privilege?' *Science*, 169 (1970), p. 427.

—, *Exploring New Ethics for Survival: The Voyage of the Spaceship Beagle* (New York: Viking Press, 1972).

—, 'Lifeboat Ethics: The Case Against Helping the Poor', *Psychology Today*, 8 (1974), pp. 38–43, 123–6.

—, 'Living on a Lifeboat', *BioScience*, 24 (1974), pp. 561–8.

—, 'Carrying Capacity as an Ethical Concept', in G. R. Lucas, Jr, and T. W. Ogletree (eds), *Lifeboat Ethics: The Moral Dilemmas of World Hunger* (New York: Harper & Row, 1976), pp. 120–37.

—, 'Ethical Implications of Carrying Capacity', in G. Hardin (ed.), *Managing the Commons* (San Francisco, CA: Freeman, 1977), pp. 112–25.

Harding, S., *Is Science Multicultural? Postcolonialisms, Feminisms, and Epistemologies* (Bloomington, IN: Indiana University Press, 1998).

Harrison, H., *Make Room! Make Room!* (1966; New York: Berkley Publishing Corporation, 1973).

Hartland, D. M., *The Story of the Space Shuttle* (London: Springer, 2004).

Harvey, D., *Justice, Nature, and the Geography of Difference* (Oxford: Blackwell, 1996).

—, *A Brief History of Neoliberalism* (Oxford: Oxford University Press, 2005).

Hawking, S., 'How can the human race survive the next hundred years?' *Yahoo!* Answers, http://answers.yahoo.com/question/?qid=20060704195516AAnrdOD [accessed 28 August 2014].

'Hawking: Mankind has 1,000 years to escape Earth', *RT Question More*, 11 April 2013, at http://rt.com/news/earth-hawking-mankind-escape-702/ [accessed 28 August 2014].

Hayles, N. K., 'Simulated Nature and Natural Simulations: Rethinking the Relation Between the Beholder and the World', in W. Cronon (ed.), *Uncommon Ground: Rethinking the Human Place in Nature* (New York and London: W. W. Norton, 1996), pp. 409–25.

Hays, S. P., *Beauty, Health and Permanence: Environmental Politics in the United States, 1955–1985* (Cambridge: Cambridge University Press, 1987).

—, 'From Conservation to Environment: Environmental Politics in the United States since World War Two', *Environmental Review*, 6 (1982), pp. 14–41.

Headrick, D. R., *The Tentacles of Progress: Technology Transfer in the Age of Imperialism, 1850–1940* (New York and Oxford: Oxford University Press, 1988).

Heims, S. J., *The Cybernetics Group* (Cambridge, MA, and London: MIT Press, 1991).

Heinrich, K., 'Das Floß der Medusa', in R. Schlesier (ed.), *Faszination des Mythos: Studien zu antiken und modernen Interpretationen* (Basel and Frankfurt: Stroemfeld and Roter Stern, 1991), pp. 335–98.

Heintz, B., and J. Huber (eds), *Mit dem Auge denken: Strategien der Sichtbarmachung in wissenschaftlichen und virtuellen Welten* (Wien and New York: Springer, 2001).

Heise, U. K., *Sense of Place and Sense of Planet: The Environmental Imagination of the Global* (Oxford and New York: Oxford University Press, 2008).

Helmreich, S., 'From Spaceship Earth to Google Ocean: Planetary Icons, Indexes, and Infrastructures', *Social Research*, 78 (2011), pp. 1211–42.

Hemingway, E., *The Old Man and the Sea* (1952; New York: Scribner, 1996).

Heppenheimer, T. A., *The Space Shuttle Decision: NASA's Search for a Reusable Space Vehicle* (Washington, DC: NASA History Office, 1999).

—, *Development of the Shuttle, 1972–1981* (Washington, DC, and London: Smithsonian Institution Press, 2002).

Hogan, T., *Mars Wars: The Rise and Fall of the Space Exploration Initiative* (Washington, DC: NASA History Division, 2007).

Höhler, S., *Luftfahrtforschung und Luftfahrtmythos: Wissenschaftliche Ballonfahrt in Deutschland, 1880–1910* (Frankfurt: Campus, 2001).

—, 'Depth Records and Ocean Volumes: Ocean Profiling by Sounding Technology, 1850–1930', *History and Technology*, 18 (2002), pp. 119–54.

—, 'The Law of Growth: How Ecology Accounted for World Population in the 20th Century', *Distinktion. Scandinavian Journal of Social Theory*, 14 (2007), pp. 45–64.

—, '"Spaceship Earth": Envisioning Human Habitats in the Environmental Age', *Bulletin of the German Historical Institute*, 42 (2008), pp. 65–85.

—, 'The Environment as a Life Support System: the Case of Biosphere 2', *History and Technology*, 26 (2010), pp. 39–58.

—, '"The Real Problem of a Spaceship Is Its People": Spaceship Earth as Ecological Science Fiction', in G. Canavan and K. S. Robinson (eds), *Green Planets: Ecology and Science Fiction* (Middletown, CT: Wesleyan University Press, 2014), pp. 99–114.

—, 'Exterritoriale Ressourcen: Die Diskussion um die Meere, die Pole und das Weltall um 1970', in I. Löhr and A. Rehling (eds), *Global Commons im 20. Jahrhundert: Entwürfe für eine globale Welt* (*Jahrbuch für Europäische Geschichte/European History Yearbook*, 15 (2014)), pp. 52–82.

Höhler, S., and F. Luks (eds), *Beam us up, Boulding! 40 Jahre 'Raumschiff Erde'*, Vereinigung für Ökologische Ökonomie – Beiträge und Berichte, 7 (2006).

Höhler, S., and R. Ziegler (eds), *Nature's Accountability* (*Science as Culture*, 19 (2010)).

Holdren, J. P., and P. R. Ehrlich (eds), *Global Ecology: Readings Toward a Rational Strategy for Man* (New York: Harcourt Brace Jovanovich, 1971).

—, 'Human Population and the Global Environment', *American Scientist*, 62 (1974), pp. 282–92.

Horn, G.-R., *The Spirit of '68: Rebellion in Western Europe and North America, 1956–1976* (Oxford: Oxford University Press, 2007).

Hünemörder, K. F., '1972 – Epochenschwelle der Umweltgeschichte?' in F.-J. Brüggemeier and J. I. Engels (eds), *Natur- und Umweltschutz nach 1945: Konzepte, Konflikte, Kompetenzen* (Frankfurt: Campus, 2005), pp. 124–44.

Hughes, D. J., *An Environmental History of the World: Humankind's Changing Role in the Community of Life* (London and New York: Routledge, 2001).

Hutchinson, G. E., 'The Biosphere', *Scientific American*, 223 (1970), pp. 45–53.

—, 'The Biosphere', in *The Biosphere*. A *Scientific American* Book (San Francisco, CA: W. H. Freeman and Company, 1970), pp. 3–11.

Ingold, T., 'Globes and Spheres: The Topology of Environmentalism', in K. Milton (ed.), *Environmentalism: The View from Anthropology* (London and New York: Routledge, 1993), pp. 29–40.

Iriye, A. (ed.), *Global Interdependence: The World after 1945* (Cambridge, MA: Belknap Press of Harvard University Press, 2013).

Jäger, M., and G. Kohn-Waechter, 'Materialien zur ökologischen Katastrophe: Das Verlassen der Erde', *Kommune* 1/1993, pp. 33–38, *Kommune*, 2 (1993), pp. 44–49, *Kommune* 3 (1993), pp. 46–51, *Kommune*, 4 (1993), pp. 50–5.

James, P. D., *The Children of Men* (1992; London: Penguin, 1994).

Jameson, F. R., *Postmodernism, or, the Cultural Logic of Late Capitalism* (1991; Durham, NC: Duke University Press, 1995).

Jamison, A., *The Making of Green Knowledge: Environmental Politics and Cultural Transfor-*

mation (Cambridge: Cambridge University Press, 2001).

Jardine, N., J. A. Secord and E. C. Spary (eds), *Cultures of Natural History* (Cambridge: Cambridge University Press, 1996).

Jasanoff, S., 'Image and Imagination: The Formation of Global Environmental Consciousness', in C. A. Miller and P. N. Edwards (eds), *Changing the Atmosphere: Expert Knowledge and Environmental Governance* (Cambridge, MA: MIT Press, 2001), pp. 309–37.

—, 'Heaven and Earth: The Politics of Environmental Images', in S. Jasanoff and M. L. Martello (eds), *Earthly Politics: Local and Global in Environmental Governance* (Cambridge, MA, and London: MIT Press, 2004), pp. 31–52.

Jewels of Light: The Stained Glass of Washington National Cathedral, eds. E. R. Crimi and D. Ney (Washington, DC: Washington National CathedralGuidebooks, 2004).

Johnson-Freese, J., *Space as a Strategic Asset* (New York: Columbia University Press, 2007).

Jørgensen, F. A., *Making a Green Machine: The Infrastructure of Beverage Container Recycling* (New Brunswick, NJ: Rutgers University Press, 2011).

Jureit, U., *Das Ordnen von Räumen: Territorium und Lebensraum im 19. und 20. Jahrhundert* (Hamburg: HIS Verlag, 2012).

Kahn, H., W. Brown and L. Martel, *The Next 200 Years: A Scenario for America and the World* (New York: Morrow, 1976).

Kahn, H., and A. J. Wiener, *The Year 2000: A Framework for Speculation on the Next Thirty-three Years* (New York: Macmillan, 1967).

Kelly, K., 'Biosphere II: An Autonomous World, Ready to Go', *Whole Earth Review*, 67 (1990), pp. 2–13.

—, 'Biosphere 2 at One', *Whole Earth Review*, 77 (1992), pp. 90–105.

—, *Out of Control: The New Biology of Machines, Social Systems and the Economic World* (Reading, MA.: Addison-Wesley, 1994).

Kennedy, J. F., Special Message to the Congress on Urgent National Needs, May 25, 1961, John F. Kennedy Presidential Library & Museum, Historical Resources, at http://www.jfklibrary.org/Asset-Viewer/Archives/JFKPOF-034–030.aspx [accessed 28 August 2014].

—, Address at Rice University on the Nation's Space Effort, September 12, 1962, NASA History Division, Key Documents in the History of Space Policy, at http://history.nasa.gov/spdocs.html [accessed 28 August 2014].

Kevles, D., 'Is the Past Prologue? Eugenics and the Human Genome Project', *Contention. Debates in Society, Culture, and Science*, 2 (1993), pp. 21–37.

Kihlstedt, F. T., 'Utopia Realized: The World's Fairs of the 1930s', in J. J. Corn (ed.), *Imagining Tomorrow: History, Technology, and the American Future* (Cambridge, MA, and London: MIT Press, 1986), pp. 97–118.

Kingsland, S. E., *Modeling Nature: Episodes in the History of Population Ecology* (1985; Chicago, IL, and London: University of Chicago Press, 1995).

—, *The Evolution of American Ecology, 1890–2000* (Baltimore, MD: Johns Hopkins University Press, 2005).

Kirk, A., 'Appropriating Technology. The Whole Earth Catalog and Counterculture Environmental Politics', *Environmental History*, 6 (2001), pp. 374–94.

Kirsch, S., *Proving Grounds: Project Plowshare and the Unrealized Dream of Nuclear Earthmoving* (New Brunswick, NJ: Rutgers University Press, 2005).

Klein, J. L., and M. S. Morgan (eds), *The Age of Economic Measurement* (Durham, NC: Duke University Press, 2001).

Klein, N., *The Shock Doctrine: The Rise of Disaster Capitalism* (New York: Henry Holt & Co,

2007).

Kleinman, D. L., *Politics on the Endless Frontier: Postwar Research Policy in the United States* (Durham, NC: Duke University Press, 1995).

Koselleck, R., *Zeitschichten: Studien zur Historik* (Frankfurt: Suhrkamp, 2000).

Krausse, J., 'Buckminster Fullers Vorschule der Synergetik', in R. Buckminster Fuller, *Bedienungsanleitung für das Raumschiff Erde und andere Schriften*, ed. J. Krausse (Amsterdam and Dresden, 1998), pp. 213–306.

Kuhn, T. S., *The Structure of Scientific Revolutions* (Chicago, IL: University of Chicago Press, 1962).

Kuhns, W., *The Post-Industrial Prophets: Interpretations of Technology* (New York: Harper and Row, 1971).

Kupper, P., 'Die "1970er Diagnose". Grundsätzliche Überlegungen zu einem Wendepunkt der Umweltgeschichte', *Archiv für Sozialgeschichte*, 43 (2003), pp. 325–48.

—, '"Weltuntergangs-Visionen aus dem Computer." Zur Geschichte der Studie "Die Grenzen des Wachstums" von 1972', in F. Uekötter and J. Hohensee (eds), *Wird Kassandra heiser? Die Geschichte falscher Ökoalarme* (Stuttgart: Franz Steiner Verlag, 2004), pp. 98–111.

—, *Creating Wilderness: A Transnational History of the Swiss National Park* (New York and Oxford: Berghahn Books, 2014).

Kwa, C., 'Local Ecologies and Global Science: Discourses and Strategies of the International Geosphere-Biosphere Programme', *Social Studies of Science*, 35 (2005), pp. 923–50.

Kylstra, P. H., and A. Meerburg, 'Jules Verne, Maury and the Ocean', in *Challenger Expedition Centenary:* Proceedings of the Second International Congress on the History of Oceanography, Edinburgh, September 12–20, 1972 (Edinburgh: Royal Society, 1972), vol. 1, pp. 243–51.

Lachmund, J., *Greening Berlin: The Co-Production of Science, Politics and Urban Nature* (Cambridge, MA, and London: MIT Press, 2013).

Lahiri-Choudhury, D. K., *Telegraphic Imperialism: Crisis and Panic in the Indian Empire, c. 1830–1920* (Basingstoke: Palgrave Macmillan, 2010).

Larsen, L. T., 'Speaking Truth to Biopower: On the genealogy of Bioeconomy', *Distinktion. Scandinavian Journal of Social Theory*, 14 (2007), pp. 9–24.

Latour, B., 'Give Me a Laboratory and I Will Raise the World', in K. Knorr-Cetina and M. Mulkay (eds), *Science Observed: Perspectives in the Social Studies of Science* (London: Sage, 1983), pp. 141–70.

—, 'Visualization and Cognition: Thinking with Eyes and Hands', *Knowledge and Society: Studies in the Sociology of Culture Past and Present*, 6 (1986), pp. 1–40.

—, *Science in Action: How to Follow Scientists and Engineers though Society* (Cambridge, MA: Harvard University Press, 1987).

—, 'Centres of Calculation', in B. Latour, *Science in Action: How to Follow Scientists and Engineers through Society* (Cambridge, MA: Harvard University Press, 1987), pp. 215–57.

—, 'Drawing Things Together', in M. Lynch and S. Woolgar (eds), *Representation in Scientific Practice* (Cambridge, MA, and London: MIT Press, 1990), pp. 19–68.

—, *We Have Never Been Modern* (1991; Cambridge, MA: Harvard University Press, 1993).

—, *Politics of Nature: How to Bring the Sciences into Democracy* (1999; Cambridge, MA, and London: Harvard University Press, 2004).

—, 'Von "Tatsachen" zu "Sachverhalten". Wie sollen die neuen kollektiven Experimente protokolliert werden?' In H. Schmidgen, P. Geimer and S. Dierig (eds) *Kultur im Experiment* (Berlin: Kadmos, 2004), pp. 17–36.

—, 'Why has Critique Run out of Steam? From Matters of Fact to Matters of Concern', *Criti-*

cal Inquiry, 30 (2004), pp. 225–48.

Launius, R. D., *Space Stations: Base Camps to the Stars* (Washington, DC, and London: Smithsonian Books, 2003).

Lefebvre, H., *La production de l'espace* (1974; Paris: Anthropos, 2000).

Legg, S., 'Foucault's Population Geographies: Classifications, Biopolitics and Governmental Spaces', *Population, Space and Place*, 11 (2005), pp. 137–56.

Lewin, R., 'Living in a Bubble', *New Scientist* (4 April 1992), pp. 12–13.

Levinson, M., *The Box: How the Shipping Container Made the World Smaller and the World Economy Bigger* (Princeton, NJ: Princeton University Press, 2006).

Levit, G. S., *Biogeochemistry, Biosphere, Noosphere: The Growth of the Theoretical System of Vladimir Ivanovich Vernadsky (1863–1945)* (Berlin: VWB, 2001).

Lewis, M., *Inventing Global Ecology* (Athens, OH: Ohio University Press, 2004).

— (ed.), *American Wilderness: A New History* (Oxford and New York: Oxford University Press, 2007).

Liebig, J. von, *Die Chemie in ihrer Anwendung auf Agricultur und Physiologie*, pt 2: *Die Naturgesetze des Feldbaues* (Braunschweig: Vieweg, 1865).

Limerick, P. N., *The Legacy of Conquest: The Unbroken Past of the American West* (New York and London: W. W. Norton, 1987).

—, 'Making the Most of Words: Verbal Activity and Western America', in W. Cronon, G. Miles and J. Gitlin (eds), *Under an Open Sky: Rethinking America's Western Past* (New York and London: W. W. Norton, 1992), pp. 167–84.

Lindenberger, T. (ed.), *Massenmedien im Kalten Krieg: Akteure, Bilder, Resonanzen* (Köln: Böhlau, 2006).

Link, J., *Versuch über den Normalismus: Wie Normalität produziert wird* (Opladen and Wiesbaden: Westdeutscher Verlag, 1999).

Linnér, B.-O., *The Return of Malthus: Environmentalism and Post-war Population-Resource Crises* (Leverburgh: White Horse Press, 2003).

Litfin, K. T., 'The Global Gaze: Environmental Remote Sensing, Epistemic Authority, and the Territorial State', in M. Hewson and T. J. Sinclair (eds), *Approaches to Global Governance Theory* (Albany, NY: SUNY Press, 1999), pp. 73–96.

Livingstone, D. N., *Putting Science in its Place: Geographies of Scientific Knowledge* (Chicago, IL: University of Chicago Press, 2003).

Logsdon, J. M., *The Decision to Go to the Moon: Project Apollo and the National Interest*, Cambridge, MA, and London: MIT Press, 1970).

Lord, D. R., *Spacelab: An International Success Story* (Washington, DC: National Aeronautics and Space Administration, Scientific and Technical Division, 1987).

Lovelock, J. E., 'Gaia as Seen Through the Atmosphere', *Atmospheric Environment,* 6 (1972), pp. 579–80.

—, *Gaia: A New Look at Life on Earth* (Oxford: Oxford University Press, 1979).

—, 'The Gaia Hypothesis', in P. Bunyard (ed.), *Gaia in Action: Science of the Living Earth* (Edinburgh: Floris, 1996), pp. 15–33.

Lovelock, J. E., and L. Margulis, 'Atmospheric Homeostasis by and for the Biosphere – the Gaia Hypothesis', *Tellus*, 26 (1974), pp. 2–10.

Luke, T. W., 'Environmental Emulations: Terraforming Technologies and the Tourist Trade at Biosphere 2', in T. W. Luke, *Ecocritique: Contesting the Politics of Nature, Economy, and Culture* (Minneapolis, MN, and London: University of Minnesota Press, 1997), pp. 95–114.

—, 'Worldwatching at the Limits of Growth', in T. W. Luke, *Ecocritique: Contesting the Poli-*

tics of Nature, Economy, and Culture (Minneapolis, MN, and London: University of Minnesota Press, 1997), pp. 75–94.

—, 'Biospheres and Technospheres: Moving from Ecology to Hyperecology with the New Class', in T. W. Luke, *Capitalism, Democracy, and Ecology: Departing from Marx* (Urbana and Chicago, IL: University of Illinois Press, 1999), pp. 59–87.

—, 'Environmentality as Green Governmentality', in É. Darier (ed.), *Discourses of the Environment* (Oxford: Blackwell, 1999), pp. 121–51.

—, 'Environmentalism as Globalization from Above and Below: Can World Watchers Truly Represent the Earth?', in P. Hayden and C. el-Ojeili (eds), *Confronting Globalization: Humanity, Justice and the Renewal of Politics* (New York: Palgrave Macmillan, 2005), pp. 154–71.

Luks, F., 'Post-normal Science and the Rhetoric of Inquiry: Deconstructing Normal Science?' *Futures*, 31 (1999), pp. 705–19.

Lutz, W., B. C. O'Neill and S. Scherbov, 'Europe's Population at a Turning Point', *Science*, 299 (2003), pp. 1991–2.

Lynch, M., and S. Woolgar (eds), *Representation in Scientific Practice* (Cambridge, MA, and London: MIT Press, 1990).

Lyotard, J.-F., *The Postmodern Condition: A Report on Knowledge* (1979; Manchester: Manchester University Press, 1984).

Lyth, P., and H. Trischler (eds), *Wiring Prometheus: Globalisation, History and Technology* (Aarhus: Aarhus University Press, 2004).

Lytle, M. H., *The Gentle Subversive: Rachel Carson, Silent Spring, and the Rise of the Environmental Movement* (New York: Oxford University Press, 2007).

MAB-UNESCO Task Force, *Criteria and Guidelines for the Choice and Establishment of Biosphere Reserves*, Program on Man and the Biosphere (MAB), Final Report (Paris: UNESCO, 1974).

McCormick, J., *The Global Environmental Movement: Reclaiming Paradise* (London: Belhaven Press, 1989).

McCray, W. P., *The Visioneers: How a Group of Elite Scientists Pursued Space Colonies, Nanotechnologies, and a Limitless Future* (Princeton, NJ, and Oxford: Princeton University Press, 2013).

McCurdy, H. E., *The Space Station Decision: Incremental Politics and Technological Choice* (Baltimore, MD: Johns Hopkins University Press, 1990).

—, *Space and the American Imagination* (Washington, DC: Smithsonian Institution Press, 1997).

McDougall, W., *... the Heavens and the Earth: A Political History of the Space Age* (1985; Baltimore, MD: Johns Hopkins University Press, 1997).

MacLeish, A., 'A Reflection: Riders on Earth Together, Brothers in Eternal Cold', *New York Times*, 25 December 1968.

McNeil, D. G. Jr, 'Population "Bomb" May Just Go "Pop!" Low Birthrates Mark a Change', *New York Times*, 6 September 2004.

McNeill, J. R., *Something New under the Sun: An Environmental History of the Twentieth Century* (London: Penguin Books, 2001).

McNeill, J. R., and C. Unger (eds), *Environmental Histories of the Cold War* (New York: Cambridge University Press, 2010).

Maier, C. S., 'Consigning the Twentieth Century to History: Alternative Narratives for the Modern Era', *American Historical Review*, 105 (2000), pp. 807–31.

—, 'Two Sorts of Crisis? The "long" 1970s in the West and the East', in H. G. Hockerts (ed.),

Koordinaten deutscher Geschichte in der Epoche des Ost-West-Konflikts (München: R. Oldenbourg Verlag, 2004), pp. 49–62.

Malthus, T., *An Essay on the Principle of Population, as It Affects the Future Improvement of Society with Remarks on the Speculations of Mr. Godwin, M. Condorcet, and Other Writers* (London, 1798).

Man and His Future: A Ciba Foundation Volume, ed. G. Wolstenholme (London: J. & A. Churchill, 1963).

Mancke, E., 'Early Modern Expansion and the Politicization of Oceanic Space', *Geographical Review*, 89 (1999), pp. 225–36.

Marcuse, H., *One-Dimensional Man: Studies in the Ideology of Advanced Industrial Society* (Boston, MA: Beacon Press, 1964).

Margulis, L., and D. Sagan, *What is Life?* (New York: Simon & Schuster, 1995).

Marino, B. D. V., and H. T. Odum, 'Biosphere 2: Introduction and Research Progress', in B. D. V. Marino and H. T. Odum (eds), *Biosphere 2: Research Past and Present* (Amsterdam: Elsevier Science, 1999), pp. 3–14.

Martel, Y., *Life of Pi* (New York: Harcourt, 2001).

Massey, D., *Space, Place, and Gender* (Cambridge: Polity Press, 1994).

—, *For Space* (London: Sage, 2005).

Matthews, W., 'Miniaturization for Space Travel', in H. D. Gilbert (ed.), *Miniaturization* (New York: Reinhold Publishing Corporation, 1961), pp. 257–69.

Maurer, E., J. Richers, M. Ruthers and C. Scheide (eds), *Soviet Space Culture: Cosmic Enthusiasm in Socialist Societies* (Houndmills, Basingstoke, and New York: Palgrave Macmillan, 2011).

Maury, M. F., *The Physical Geography of the Sea* (1855; New York and London: Harper and Sampson Low, 1859).

Mawer, G. A., *South by Northwest: The Magnetic Crusade and the Contest for Antarctica* (Edinburgh: Birlinn, 2006).

Meadows, D. H., D. L. Meadows, J. Randers, W. W. Behrens III, *The Limits to Growth: A Report for the Club of Rome's Project on the Predicament of Mankind* (New York: Universe Books, 1972).

Meadows, D. H., D. L. Meadows and J. Randers, *Beyond the Limits* (Post Mills, VT: Chelsea Green Publishing, 1992).

Meadows, D. H., J. Randers and D. Meadows, *Limits to Growth: The 30-Year Update* (White River Junction, VT: Chelsea Green Publishing Company, 2004).

Melosi, M. V., *Atomic Age America* (Boston, MA: Pearson, 2012).

Melville, H., *Moby-Dick, or, The Whale* (1851; New York and London: Penguin Books, 1992).

Mendelsohn, E., 'The Politics of Pessimism: Science and Technology Circa 1968', in Y. Ezrahi, E. Mendelsohn and H. Segal (eds), *Technology, Pessimism, and Postmodernism* (Dordrecht: Kluwer, 1994), pp. 151–73.

Merchant, C., *The Death of Nature: Women, Ecology, and the Scientific Revolution* (San Francisco, CA: Harper & Row, 1980).

—, *Reinventing Eden: The Fate of Nature in Western Culture* (New York and London: Routledge, 2004).

Miller, J. G., and J. L. Miller, 'The Earth as System', *Behavioral Science*, 27 (1982), pp. 303–22.

Miller, R., *The Dream Machines: An Illustrated History of the Spaceship in Art, Science and Literature* (Malabar, FL: Krieger, 1993).

Mitman, G., M. Murphy and C. Sellers (eds), *Landscapes of Exposure: Knowledge and Illness in Modern Environments* (*Osiris*, 19 (2004)).

Mitsch, W. J., 'Preface: Biosphere 2 – The Special Issue', in B. D. V. Marino and H.T. Odum (eds), *Biosphere 2: Research Past and Present* (Amsterdam: Elsevier Science, 1999), pp. 1–2.

Mol, A. P. J., *Globalization and Environmental Reform: The Ecological Modernization of the Global Economy* (Cambridge, MA: MIT Press, 2001).

Moll, P., *From Scarcity to Sustainability: Futures Studies and the Environment. The Role of the Club of Rome* (Frankfurt: Peter Lang Verlag, 1991).

Monod, J., *Chance and Necessity: An Essay on the Natural Philosophy of Modern Biology* (1970; New York: Vintage Books, 1971).

Morgan, S. P., 'Is Low Fertility a Twenty-First-Century Demographic Crisis?' *Demography*, 40 (2003), pp. 589–603.

Moses, R., 'The Fashioning of a Fair', in *New York World's Fair 1964/1965: Official Souvenir Book* (New York: Time Inc., 1964), pp. 15–19.

Myrdal, A., and P. Vincent, *Are there Too Many People?* (New York: Manhattan Publishing Company, 1950).

Murphy, M., *Sick Building Syndrome and the Problem of Uncertainty: Environmental Politics, Technoscience, and Women Workers* (Durham, NC: Duke University Press, 2006).

Murphy, P. C., *What a Book Can Do: The Publication and Reception of Silent Spring* (Amherst and Boston, MA: University of Massachusetts Press, 2005).

Næss, P., 'Live and Let Die: The Tragedy of Hardin's Social Darwinism', *Journal of Environmental Policy & Planning*, 6 (2004), pp. 19–34.

'NASA Leadership and America's Future in Space'. A Report to the Administrator by Dr. Sally K. Ride, August 1987. Key Documents in the History of Space Policy, National Aeronautics and Space Administration, at http://www.hq.nasa.gov/office/pao/History/sp-docs.html [accessed 28 August 2014].

Natter, W., 'Friedrich Ratzel's Spatial Turn: Identities of Disciplinary Space and its Borders Between the Anthropo- and Political Geography of Germany and the United States', in H. van Houtum, O. Kramsch and W. Zierhofer (eds), *B/ordering Space* (Aldershot: Ashgate, 2005), pp. 171–85.

Natter, W., and J. P. Jones, 'Identity, Space, and other Uncertainties', in G. Benko and U. Strohmayer (eds), *Space and Social Theory: Interpreting Modernity and Postmodernity* (Oxford: Basil Blackwell, 1997), pp. 141–61.

Nayder, L., 'Sailing Ships and Steamers, Angels and Whores: History and Gender in Joseph Conrad's Maritime Fiction', in M. S. Creighton and L. Norling (eds), *Iron Men, Wooden Women: Gender and Seafaring in the Atlantic World, 1700–1920* (Baltimore, MD, and London: Johns Hopkins University Press, 1996), pp. 189–203.

Neurath, P., *From Malthus to the Club of Rome and Back: Problems of Limits to Growth, Population Control, and Migrations* (Armonk, NY: M. E. Sharpe, 1994).

New York World's Fair 1964/1965: Official Guide (New York: Time Inc., 1964).

Nicholson, M., *The Environmental Revolution: A Guide for the New Masters of the World* (New York: McGraw-Hill, 1970).

Nobles, G. H., 'Straight Lines and Stability: Mapping the Political Order of the Anglo-American Frontier', *Journal of American History*, 80 (1993), pp. 9–35.

Nolan, W. F., and G. C. Johnson, *Logan's Run* (New York: Dial Press, 1967).

Nye, D. E., *America as Second Creation. Technology and Narratives of New Beginnings* (Cambridge, MA: MIT Press, 2004).

Odum, E. P., *Fundamentals of Ecology* (Philadelphia, PA: W. B. Saunders, 1953).

—, 'Cost of Living in Domed Cities', *Nature*, 382 (1996), p. 18.

Odum, H. T., *Environment, Power, and Society* (New York: Wiley-Interscience, 1971).

Ogilvie, S. A., and S. Miller, *Refuge Denied: The St. Louis Passengers and the Holocaust* (Madison, WI: University of Wisconsin Press, 2006).

O'Neill, G. K., *The High Frontier: Human Colonies in Space* (New York: Morrow, 1976).

One Planet, Many People: Atlas of Our Changing Environment (Nairobi: UNEP, 2005).

Osborn, F., *Our Plundered Planet* (1948; Boston, MA: Little, Brown & Co., 1950).

—, *The Limits of the Earth* (Boston, MA: Little, Brown & Co., 1953).

—, *Our Crowded Planet: Essays on the Pressures of Population* (Garden City, NY: Doubleday, 1962).

Osterhammel, J., *Die Verwandlung der Welt: Eine Geschichte des 19. Jahrhunderts* (München: Beck, 2009).

Our Common Future, World Commission on Environment and Development (Oxford and New York: Oxford University Press, 1987).

Paddock, W., and P. Paddock, *Famine – 1975!* (1967; London: Weidenfeld & Nicolson, 1968).

Park, D., *The Grand Contraption: The World as Myth, Number, and Chance* (Princeton, NJ: Princeton University Press, 2005).

Pawley, E., *'The Balance-Sheet of Nature': Calculating the New York Farm, 1820–1860* (PhD Dissertation Dickinson College, ProQuest, UMI Dissertation Publishing, 2011).

Pearl, R., *The Biology of Population Growth* (1925; New York: Alfred A. Knopf, 1930).

Penley, C., *NASA/Trek: Popular Science and Sex in America* (London and New York: Verso, 1997).

Pias, C. (ed.), *Cybernetics/Kybernetik: The Macy-Conferences 1946–1953* (Zürich and Berlin: Diaphanes, 2003–4).

Pickering, A., *The Cybernetic Brain: Sketches of Another Future* (Chicago, IL: University of Chicago Press, 2009).

'Pioneering the Space Frontier.' The Report of the National Commission on Space, May 1986. Key Documents in the History of Space Policy, National Aeronautics and Space Administration, at http://www.hq.nasa.gov/office/pao/History/spdocs.html [accessed 28 August 2014].

Piotrow, P. T., *World Population Crisis: The United States Response* (New York: Praeger, 1973).

Pörksen, U., 'Logos, Kurven, Visiotype', in U. Gerhard, J. Link and E. Schulte-Holtey (eds), *Infografiken, Medien, Normalisierung: Zur Kartographie politisch-sozialer Landschaften* (Heidelberg: Synchron, 2001), pp. 63–76.

Pollard, W. G., *Man on a Space Ship: The Meaning of the Twentieth Century Revolution and the Status of Man in the Twenty-first and After* (Claremont, CA: Claremont Colleges, 1967).

Polunin, N., 'Our Use of "Biosphere", "Ecosystem" and now "Ecobiome"', *Environmental Conservation*, 11 (1984), p. 198.

Porter, R., and M. Teich (eds), *The Scientific Revolution in National Context* (Cambridge and New York: Cambridge University Press, 1992).

Porter, T. M., 'The Mathematics of Society: Variation and Error in Quetelet's Statistics', *British Journal for the History of Science*, 18 (1985), pp. 51–69.

—, *The Rise of Statistical Thinking, 1820–1900* (Princeton, NJ: Princeton University Press, 1986).

—, *Trust in Numbers: The Pursuit of Objectivity in Science and Public Life* (Princeton, NJ: Princeton University Press, 1995).

—, *Karl Pearson: The Scientific Life in a Statistical Age* (Princeton, NJ: Princeton University Press, 2004).

—, 'The Culture of Quantification and the History of Public Reason', *Journal of the History of Economic Thought*, 26 (2004), pp. 165–77.

Poynter, J., *The Human Experiment: Two Years and Twenty Minutes inside Biosphere 2* (New York: Thunder's Mouth Press, 2006).

Pratt, M. L., *Imperial Eyes: Travel Writing and Transculturation* (London and New York: Routledge, 1992).

'President Bush's Remarks on the Twentieth Anniversary of the Apollo 11 Moon Landing (his space exploration initiative speech), October 4, 1989', Key Documents in the History of Space Policy, National Aeronautics and Space Administration, at http://www. hq.nasa.gov/office/pao/History/spdocs.html [accessed 28 August 2014].

'President Nixon's Announcement on the Development of the Space Shuttle, January 5, 1972', Key Documents in the History of Space Policy, National Aeronautics and Space Administration, at http://www.hq.nasa.gov/office/pao/History/spdocs.html [accessed 28 August 2014].

Quételet, A., *A Treatise on Man and the Development of his Faculties*. Facsimile Reproduction of the English Translation of 1842 (1835; New York: Burt Franklin, 1968).

Radkau, J., *Nature and Power: A Global History of the Environment* (2002; Cambridge and New York: Cambridge University Press, 2008).

Ramsden, E., 'Carving up Population Science: Eugenics, Demography and the Controversy over the "Biological Law" of Population Growth', *Social Studies of Science*, 32 (2002), pp. 857–99.

—, 'Confronting the Stigma of Eugenics: Genetics, Demography and the Problems of Population', *Social Studies of Science*, 39 (2009), pp. 853–84.

Rau, S., *Räume: Konzepte, Wahrnehmungen, Nutzungen* (Frankfurt: Campus, 2013).

Raum: Ein interdisziplinäres Handbuch, ed. S. Günzel (Stuttgart and Weimar: Verlag J. B. Metzler, 2010).

Reider, R., *Dreaming the Biosphere: The Theater of All Possibilities* (Albuquerque, NM: University of New Mexico Press, 2009).

Richards, T., *The Imperial Archive: Knowledge and the Fantasy of Empire* (London and New York: Verso, 1993).

Rifkin, J., *Algeny: A New Word – A New World* (Harmondsworth: Penguin, 1983).

—, *Biosphere Politics: A New Consciousness for a New Century* (New York: Crown, 1991).

Robertson, T., *The Malthusian Moment: Global Population Growth and the Birth of American Environmentalism* (New Brunswick, NJ: Rutgers University Press, 2012).

Rome, A., *The Genius of Earth Day: How a 1970 Teach-In Unexpectedly Made the First Green Generation* (New York: Hill & Wang, 2013).

Rosenberg, E. S. (ed.), *A World Connecting, 1870–1945* (Cambridge, MA: Belknap Press of Harvard University Press, 2012).

Ross, A., *Strange Weather: Culture, Science and Technology in the Age of Limits* (London: Verso, 1991).

Ross, E. A., *Standing Room Only?* (New York: Century, 1927).

Rothman, H. K., *Saving the Planet: The American Response to the Environment in the Twentieth Century* (Chicago, IL: Ivan R. Dee, 2000).

Rozwadowski, H. M., 'Small World: Forging a Scientific Maritime Culture for Oceanography', *Isis*, 87 (1996), pp. 409–29.

—, *Fathoming the Ocean: The Discovery and Exploration of the Deep Sea* (Cambridge, MA, and London: Belknap Press of Harvard University Press, 2005).

Russell, E., *War and Nature: Fighting Humans and Insects with Chemicals from World War I to*

Silent Spring (New York: Cambridge University Press, 2001).

Sachs, W., 'Natur als System. Vorläufiges zur Kritik der Ökologie', *Scheidewege: Jahresschrift für skeptisches Denken*, 21 (1991/2), pp. 83–97.

—, 'One World', in W. Sachs (ed.), *The Development Dictionary: A Guide to Knowledge as Power* (London: Zed Books, 1992), pp. 102–15.

—, 'Zählen oder Erzählen? Natur- und geisteswissenschaftliche Argumente in der Studie "Zukunftsfähiges Deutschland"', *Wechselwirkung*, 17 (1995), pp. 20–5.

—, 'Astronautenblick – Über die Versuchung zur Weltsteuerung in der Ökologie', in *Jahrbuch Ökologie 1999* (München: C. H. Beck, 1998), pp. 199–206.

—, *Planet Dialectics: Explorations in Environment and Development* (London: Zed Books, 1999).

—, 'Is small still beautiful? E. F. Schumacher im Zeitalter der grenzenlosen Mega-Ökonomie', in *Re-Vision: Nachdenken über ökologische Vordenker* (*Politische Ökologie*, 24 (2006)), pp. 24–6.

— (ed.), *Der Planet als Patient: Über die Widersprüche globaler Umweltpolitik* (Basel: Birkhäuser, 1994).

Sagan, C., *Pale Blue Dot: A Vision of the Human Future in Space* (New York: Random House, 1994).

Sagan, D., *Biospheres: Metamorphosis of Planet Earth* (New York et al.: Bantam Books, 1990).

Sahm, P. R., H. Rahmann, H. J. Blome and G. P. Thiele (eds), *Homo spaciens: Der Mensch im Kosmos* (Hamburg: Discorsi, 2005).

Salas, R. M., *International Population Assistance: The First Decade. A Look at the Concepts and Politics Which have Guided the UNFPA in its First Ten Years* (New York: Pergamon Press, 1979).

Salisbury, F. B., 'Joseph I. Gitelson and the Bios-3 project', *Life Support & Biosphere Science. International Journal of Earth Space*, 1 (1994), pp. 69–70.

Samson, P. R., and D. Pitt (eds), *The Biosphere and the Noosphere Reader: Global Environment, Society and Change* (London and New York: Routledge, 1999).

Sanger, M. (ed.), *Proceedings of the World Population Conference, held at the Salle Centrale, Geneva, August 29th to September 3rd, 1927* (London: Edward Arnold & Co., 1927).

Saraiva, T., 'Breeding Europe: Crop Diversity, Gene Banks, and Commoners', in N. Disco and E. Kranakis (eds), *Cosmopolitan Commons: Sharing Resources and Risks Across Borders* (Cambridge, MA: MIT Press, 2013), pp. 185–212.

Sax, K., *Standing Room Only: The World's Exploding Population* (1955; Boston, MA: Beacon Press, 1960).

Schellnhuber, H. J., '"Earth System" Analysis and the Second Copernican Revolution', *Nature*, 402 (1999), pp. C19–C23.

Schiebinger, L., *Nature's Body: Gender in the Making of Modern Science* (New Brunswick, NJ: Rutgers University Press, 1993).

—, *Plants and Empire: Colonial Bioprospecting in the Atlantic World* (Cambridge, MA: Harvard University Press, 2004).

Schmidt, A., *Die Gelehrtenrepublik: Kurzroman aus den Roßbreiten* (1957; Frankfurt: Fischer, 2004).

Schmidt-Gernig, A., 'The Cybernetic Society: Western Future Studies of the 1960s and 1970s and their Predictions for the Year 2000', in R. Layard and R. N. Cooper (eds), *What the Future Holds: Insights from Social Science* (2002; Cambridge, MA: MIT Press, 2003), pp. 233–59.

Schoijet, M., '*Limits to Growth* and the Rise of Catastrophism', *Environmental History*, 4

(1999), pp. 515–30.

Schröder, I., *Das Wissen von der ganzen Welt: Globale Geographien und räumliche Ordnungen Afrikas und Europas, 1790–1870* (Paderborn: Schöningh, 2011).

Schröder, I., and S. Höhler (eds), *Welt-Räume: Geschichte, Geographie und Globalisierung seit 1900* (Frankfurt: Campus, 2005).

—, 'Für eine Geschichte der Orte und Räume im globalen Zeitalter', in I. Schröder and S. Höhler (eds), *Welt-Räume: Geschichte, Geographie und Globalisierung seit 1900* (Frankfurt: Campus, 2005), pp. 303–13.

Schrödinger, E., *What is Life? The Physical Aspect of the Living Cell.* With *Mind and Matter* and Autobiographical Sketches (Cambridge and New York: Cambridge University Press, 1992).

Schumacher, E. F., *Small Is Beautiful: A Study of Economics as if People Mattered* (London: Blond Briggs, 1973).

Scott, J. C., *Seeing Like a State: How Certain Schemes to Improve the Human Condition Have Failed* (New Haven, CT: Yale University Press, 1998).

Scott-Heron, G., 'Whitey on the Moon', *The Revolution Will Not Be Televised* (Flying Dutchman/RCA, 1974), http://www.gilscottheron.com/lywhitey.html [accessed 28 August 2014].

Seamans, R. C. Jr, *Project Apollo: The Tough Decisions* (Washington, DC: NASA History Division, 2005).

Shapin, S., *The Scientific Revolution* (Chicago, IL: University of Chicago Press, 1996).

Shayler, D. J., *Skylab: America's Space Station* (London: Springer, 2001).

Siddiqi, A. A., *The Soviet Space Race with Apollo* (Gainesville, FL: University Press of Florida, 2003).

'Sim-Bio 2: Investigate Biosphere 2 by Modeling and Designing Experiments', *BioQuest Curriculum Consortium*, at http://bioquest.org/simbio2.html [accessed 28 August 2014].

Simon, J. L., *The Ultimate Resource* (Princeton, NJ: Princeton University Press, 1981).

Sloterdijk, P., *Sphären*, Vol. 2: *Globen (Makrosphärologie)* (Frankfurt: Suhrkamp, 1999).

Smil, Vaclav, *The Earth's Biosphere: Evolution, Dynamics, and Change* (Cambridge, MA, and London: MIT Press, 2003).

—, *Harvesting the Biosphere: What We Have Taken From Nature* (Cambridge, MA: MIT Press, 2013).

Smith, H. N., *Virgin Land: The American West as Symbol and Myth* (1950; New York: Vintage Books, 1957).

Smith, M., 'Against Ecological Sovereignty: Agamben, Politics and Globalisation', *Environmental Politics*, 18 (2009), pp. 99–116.

Sohn, W., and H. Mehrtens (eds), *Normalität und Abweichung: Studien zur Geschichte und Theorie der Normalisierungsgesellschaft* (Opladen, Wiesbaden: Westdeutscher Verlag, 1999).

Soja, E. W., *Postmodern Geographies: The Reassertion of Space in Critical Social Theory* (1989; London: Verso, 1999).

Sokal, A. D., and J. Bricmont, *Fashionable Nonsense: Postmodern Intellectuals' Abuse of Science* (New York: Picador, 1998).

Sorrenson, R., 'The Ship as a Scientific Instrument in the Eighteenth Century', in H. Kuklick and R. E. Kohler (eds), *Science in the Field* (*Osiris*, 11 (1996)), pp. 221–36.

Stanek, B., *Raumfahrt Lexikon* (Bern and Stuttgart: Hallwag, 1983).

Stegner, W., *The American West as Living Space* (Ann Arbor, MI: University of Michigan Press, 1987).

Sueß, E., *Die Entstehung der Alpen* (Wien: Braunmüller, 1875).

—, *Das Antlitz der Erde*, 3 vols (Prag, Wien and Leipzig: F. Tempsky and G. Freytag, 1885–1909).

Tansley, A. G., 'The Use and Abuse of Vegetational Concepts and Terms', *Ecology*, 16 (1935), pp. 284–307.

Taylor, G. R., *The Biological Time-Bomb* (London: Thames & Hudson, 1968).

—, *The Doomsdaybook* (London: Thames & Hudson, 1970).

Taylor, P. J., and F. H. Buttel, 'How Do We Know We Have Global Environmental Problems? Science and the Globalization of Environmental Discourse', *Geoforum*, 23 (1992), pp. 405–16.

Taylor, P., 'Technocratic Optimism, H. T. Odum, and the Partial Transformation of Ecological Metaphor after World War II', *Journal of the History of Biology*, 21 (1988), pp. 213–44.

Tenbruck, F. H., 'The Dream of a Secular Ecumene: The Meaning and Limits of Policies of Development', in M. Featherstone (ed.), *Global Culture: Nationalism, Globalization and Modernity* (London: Sage, 1990), pp. 193–206.

'The Apollo 8 Christmas Eve Broadcast', National Space Science Data Center, National Aeronautics and Space Administration, at http://nssdc.gsfc.nasa.gov/planetary/lunar/apollo8_xmas.html [accessed 28 August 2014].

'The Post-Apollo Space Program: Directions for the Future, September 1969', National Aeronautics and Space Administration, at http://www.hq.nasa.gov/office/pao/History/taskgrp.html [accessed 28 August 2014].

The Biosphere (San Francisco, CA: W. H. Freeman and Company, 1970).

The Far Side, by Gary Larson (Kansas City and New York: Andrews and McMeel, 1982).

Thomas, G., and M. Morgan Witts, *Voyage of the Damned* (New York: Stein and Day, 1974).

Tilley, H., *Africa as a Living Laboratory: Empire, Development, and the Problem of Scientific Knowledge, 1870–1950* (Chicago, IL: University of Chicago Press, 2011).

Toffler, A., *Future Shock* (New York: Random House, 1970).

Traven, B., *The Death Ship: The Story of an American Sailor* (1926; London: Chatto & Windus, 1934).

Tucker, R. P., and E. Russell (eds), *Natural Enemy, Natural Ally: Toward an Environmental History of War* (Corvallis, OR: Oregon State University Press, 2004).

Turner, F., *From Counterculture to Cyberculture: Stewart Brand, the Whole Earth Network, and the Rise of Digital Utopianism* (Chicago, IL: University of Chicago Press, 2006).

Turner, F. J., *The Frontier in American History* (New York: Henry Holt & Co., 1921).

Twain, M., *The Adventures of Huckleberry Finn*, adapted by Matthew Francis (1884; London and New York: S. French, 1998).

Use and Conservation of the Biosphere, Proceedings of the Intergovernmental Conference of Experts on the Scientific Basis for Rational Use and Conservation of the Resources of the Biosphere, Paris, 4–13 September 1968 (Liège: Unesco, 1970).

van Dieren, W. (ed.), *Taking Nature into Account: A Report to the Club of Rome* (New York: Springer, 1995).

Veggeberg, S., 'Escape from Biosphere 2', *New Scientist*, 25 September 1993, pp. 22–4.

Vernadsky, V. I., *The Biosphere* (1926; New York and Heidelberg: Copernicus and Springer, 1998).

Verne, J., *20,000 Leagues Under the Sea* (1870; New York: Penguin Books and Signet Classics, 2001).

—, *Around the World in Eighty Days* (1873; New York: Penguin Books and Signet Classics,

2005).

Völker-Rasor, A., and W. Schmale (eds), *MythenMächte – Mythen als Argument* (Berlin: Berlin Verlag Arno Spitz, 1998).

Wackernagel, M., and W. Rees, *Our Ecological Footprint: Reducing Human Impact on the Earth* (Gabriola Island, BC: New Society Publishers, 1996).

Wang, Z., *In Sputnik's Shadow: The President's Science Advisory Committee and Cold War America* (New Brunswick, NJ: Rutgers University Press, 2008).

Ward, B., *Spaceship Earth* (New York: Columbia University Press, 1966).

Ward, B., and R. Dubos, *Only One Earth: The Care and Maintenance of a Small Planet*. An Unofficial Report Commissioned by the Secretary-General of the United Nations Conference on the Human Environment, Prepared with the Assistance of a 152-Member Committee of Corresponding Consultants in 58 Countries (New York: Norton, 1972).

Warshall, P., 'The Ecosphere: Introducing an Einsteinian Ecology', *Whole Earth Review*, 46 (1985), pp. 28–31.

—, 'Lessons from Biosphere 2: Ecodesign, Surprises, and the Humility of Gaian Thought', *Whole Earth Review*, 89 (1996), pp. 22–7.

Washington National Cathedral: Guidebook to the Cathedral (Washington, DC: Washington National Cathedral Guidebooks, 2004).

Webb, W. P., *Divided We Stand: The Crisis of a Frontierless Democracy* (New York: Farrar & Rinehart, 1937).

—, *The Great Frontier* (Boston, MA: Houghton Mifflin Co., 1952).

Weber, B., 'Ubi Caelum Terrae Se Coniungit: Ein altertümlicher Aufriß des Weltgebäudes von Camille Flammarion', *Gutenberg-Jahrbuch* (1975), pp. 381–408.

Weiss, E. B., *In Fairness to Future Generations: International Law, Common Patrimony, and Intergenerational Equity* (Tokyo and Dobbs Ferry, NY: Transnational Publishers, 1989).

Weissman, S., 'Die Bevölkerungsbombe ist ein Rockefeller-Baby', in H. M. Enzensberger and K. M. Michel (eds), *Ökologie und Politik oder Die Zukunft der Industrialisierung* (Berlin: Kursbuch/Rotbuch Verlag, 1973), pp. 81–94.

Werth, K., *Ersatzkrieg im Weltraum: Das US-Raumfahrtprogramm in der Öffentlichkeit der 1960er Jahre* (Frankfurt: Campus, 2006).

Wertheim, M., *The Pearly Gates of Cyberspace: A History of Space from Dante to the Internet* (New York and London: W. W. Norton, 1999).

Wharton, R. A. Jr., D. T. Smernoff and M. M. Averner, 'Algae in Space', in C. A. Lembi and J. R. Waaland (eds), *Algae and Human Affairs* (Cambridge: Cambridge University Press, 1988).

Whatmore, S., *Hybrid Geographies: Natures of Cultural Spaces* (London: Sage, 2002).

Williams, G., *Voyages of Delusion: The Quest for the Northwest Passage* (New Haven, CT: Yale University Press, 2003).

Wöbse, A.-K., *Weltnaturschutz: Umweltdiplomatie in Völkerbund und Vereinten Nationen, 1920–1950* (Frankfurt: Campus, 2011).

Woodward, D., 'The Image of the Spherical Earth', *Perspecta. The Yale Architectural Journal*, 25 (1989), pp. 2–15.

Woodwell, G. M., 'The Energy Cycle of the Biosphere', in *The Biosphere. A Scientific American Book* (San Francisco, CA: W. H. Freeman and Company, 1970), pp. 26–36.

Wormbs, N. 'Eyes on the Ice: Satellite Remote Sensing and the Narratives of Visualized Data', in M. Christensen, A. Nilsson and N. Wormbs (eds), *Media and the Politics of Arctic Climate Change: When the Ice Breaks* (Basingstoke: Palgrave Macmillan, 2013).

Worster, D., *Nature's Economy: A History of Ecological Ideas* (Cambridge, MA, and New York:

Cambridge University Press, 1977).

Wrobel, D. M., *The End of American Exceptionalism: Frontier Anxiety from the Old West to the New Deal* (Lawrence, KS: University Press of Kansas, 1993).

Zabel, B., P. Hawes, H. Stuart and B. D. V. Marino, 'Construction and Engineering of a Created Environment: Overview of the Biosphere 2 Closed System', in B. D. V. Marino and H. T. Odum (eds), *Biosphere 2: Research Past and Present* (Amsterdam: Elsevier Science, 1999), pp. 43–63.

Zachary, G. P., *Endless Frontier: Vannevar Bush, Engineer of the American Century* (Cambridge, MA, and London: MIT Press, 1999).

FILMOGRAPHY

2001: A Space Odyssey, directed by Stanley Kubrick, starring Keir Dullea, Gary Lockwood and William Sylvester (USA: Metro-Goldwyn-Mayer, 1968).

Children of Men, directed by Alfonso Cuarón, starring Clive Owen, Julianne Moore and Michael Caine (USA/GB: Universal, 2006) (DVD includes *The Possibility of Hope*, a documentary film by Alfonso Cuarón (USA: Universal Studios, 2006).

Das Boot, directed by Wolfgang Petersen, starring Jürgen Prochnow, Herbert Grönemeyer and Klaus Wennemann (Germany: Bavaria Atelier/Radiant Film/WDR, 1981).

Logan's Run, directed by Michael Anderson, starring Michael York, Jenny Agutter and Peter Ustinov (USA: Warner Brothers/Metro-Goldwyn-Mayer, 1976).

Silent Running, directed by Douglas Trumbull, starring Bruce Dern (USA: Universal, 1972).

Soylent Green, directed by Richard Fleischer, starring Charlton Heston, Leigh Taylor Young and Edward G. Robinson (USA: Warner Brothers/Metro-Goldwyn-Mayer, 1973).

The Lifeboat, directed by Alfred Hitchcock, starring Tallulah Bankhead and William Bendix (USA: 20th Century Fox, 1944).

The Noah, directed by Daniel Bourla, starring Robert Strauss (USA, 1974).

Total Recall, directed by Paul Verhoeven, starring Arnold Schwarzenegger, Sharon Stone and Rachel Ticotin (USA: TriStar Pictures, 1990).

WALL•E, directed by Andrew Stanton (USA: Walt Disney Pictures/Pixar Animation Studios, 2008).

Zardoz, directed by John Boorman, starring Sean Connery, Charlotte Rampling and John Alderton (USA: 20th Century Fox, 1974).

Z.P.G. Zero Population Growth, directed by Michael Campus, starring Oliver Reed and Geraldine Chaplin (USA: Paramount, 1971).

NOTES

1 Capacity: Environment in a Century of Space

1. D. H. Meadows, D. L. Meadows, J. Randers, W. W. Behrens III, *The Limits to Growth: A Report for the Club of Rome's Project on the Predicament of Mankind* (New York: Universe Books, 1972).

2. M. Foucault, 'Of Other Spaces' [orig. French 'Des Espaces Autres', lecture manuscript 1967], *Diacritics*, 16 (1986), pp. 22–7, on p. 22.

3. E. Goldsmith, R. Allen, M. Allaby, J. Davoll and S. Lawrence, *A Blueprint for Survival* (London: Tom Stacey, 1972).

4. On new global history writing see C. A. Bayly, *The Birth of the Modern World: Global Connections and Comparisons, 1780–1914* (Oxford: Blackwell, 2004); A. Iriye (ed.), *Global Interdependence: The World after 1945* (Cambridge, MA: Belknap Press of Harvard University Press, 2013); J. Osterhammel, *Die Verwandlung der Welt: Eine Geschichte des 19. Jahrhunderts* (München: Beck, 2009); E. S. Rosenberg (ed.), *A World Connecting, 1870–1945* (Cambridge, MA: Belknap Press of Harvard University Press, 2012).

5. S. Höhler, 'Exterritoriale Ressourcen: Die Diskussion um die Meere, die Pole und das Weltall um 1970', in I. Löhr and A. Rehling (eds), *Global Commons im 20. Jahrhundert: Entwürfe für eine globale Welt* [*Jahrbuch für Europäische Geschichte/European History Yearbook*, 15 (2014), pp. 53–82.

6. Questions on the evidence of space and place have long been the domains of poststructuralist social sciences, cultural studies, postcolonial studies and critical geography; for example, A. Appadurai, *Modernity at Large: Cultural Dimensions of Globalization* (Minneapolis, MN: University of Minnesota Press, 1996); M. Crang, and N. Thrift (eds), *Thinking Space* (New York and London: Routledge, 2000); D. Gregory, *Power, Knowledge and Geography: An Introduction to Geographic Thought and Practice* (Oxford: Blackwell, 2006); D. Harvey, *Justice, Nature, and the Geography of Difference* (Oxford: Blackwell, 1996); H. Lefebvre, *La production de l'espace* (1974; Paris: Anthropos, 2000); D. Massey, *For Space* (London: Sage, 2005); E. W. Soja, *Postmodern Geographies: The Reassertion of Space in Critical Social Theory* (1989; London: Verso, 1999); S. Whatmore, *Hybrid Geographies: Natures of Cultural Spaces* (London: Sage, 2002). In the past decade, space and place have become objects of historical reflection. For literature on the programmatic endeavour of what might be called the 'spatial turn' in history see, for example, D. Chakrabarty, *Provincializing Europe: Postcolonial Thought and Historical Difference* (Princeton, NJ: Princeton University Press, 2000); I. Schröder

and S. Höhler (eds), *Welt-Räume: Geschichte, Geographie und Globalisierung seit 1900* (Frankfurt: Campus, 2005); I. Schröder, *Das Wissen von der ganzen Welt: Globale Geographien und räumliche Ordnungen Afrikas und Europas, 1790–1870* (Paderborn: Schöningh, 2011).

7. Foucault, 'Of Other Spaces', p. 23.

8. J. P. Holdren and P. R. Ehrlich, 'Human Population and the Global Environment', *American Scientist*, 62 (1974), pp. 282–92, on p. 290.

9. For linking imperialism and technological internationalism see, for instance, M. H. Geyer and J. Paulmann (eds), *The Mechanics of Internationalism: Culture, Society and Politics from the 1840s to the First World War* (Oxford: Oxford University Press, 2001); P. Lyth and H. Trischler (eds), *Wiring Prometheus: Globalisation, History and Technology* (Aarhus: Aarhus University Press, 2004). On science and imperialism see, for example, A. W. Crosby, *Ecological Imperialism: The Biological Expansion of Europe, 900–1900* (New York: Cambridge University Press, 1986); M. H. Edney, *Mapping an Empire: The Geographical Construction of British India, 1765–1843* (Chicago, IL: University of Chicago Press, 1997); B. Etemad, *Possessing the World: Taking the Measurements of Colonisation from the 18th to the 20th Century* (New York and Oxford: Berghahn Books, 2007); F. Fan, *British Naturalists in Qing China: Science, Empire and Cultural Encounter* (Cambridge, MA, and London: Harvard University Press, 2004); D. N. Livingstone, *Putting Science in its Place: Geographies of Scientific Knowledge* (Chicago, IL: University of Chicago Press, 2003); L. Schiebinger, *Plants and Empire: Colonial Bioprospecting in the Atlantic World* (Cambridge, MA: Harvard University Press, 2004); H. Tilley, *Africa as a Living Laboratory: Empire, Development, and the Problem of Scientific Knowledge, 1870–1950* (Chicago, IL: University of Chicago Press, 2011).

10. C. S. Maier, 'Consigning the Twentieth Century to History: Alternative Narratives for the Modern Era', *American Historical Review*, 105 (2000), pp. 807–31, on p. 807.

11. F. Osborn, *The Limits of the Earth* (Boston, MA: Little, Brown & Co., 1953), p. 78, here on the voyages of Magellan.

12. W. P. Webb, *The Great Frontier* (Boston, MA: Houghton Mifflin Co., 1952).

13. J. Brunhes, *Les Limites de notre cage*, Discours à l'occasion de l'inauguration solennelle des cours universitaires, le 15 novembre 1909 (Fribourg, 1911).

14. The concept of *Lebensraum*, the German term framing the political debate on living space in the early twentieth century, shares common ecological origins and lines of development with the concept of biosphere as discussed in this book. On the geopolitical meaning of *Lebensraum* see U. Jureit, *Das Ordnen von Räumen: Territorium und Lebensraum im 19. und 20. Jahrhundert* (Hamburg: HIS Verlag, 2012).

15. R. Moses, 'The Fashioning of a Fair', in *New York World's Fair 1964/1965: Official Souvenir Book* (New York: Time Inc., 1964), pp. 15–19. Robert Moses was New York City's Park Commissioner. For further information on the 1964/5 New York's World Fair and Unisphere, see *New York World's Fair 1964/1965: Official Guide* (New York: Time Inc., 1964), especially pp. 90, 178–80, and I. Schröder and S. Höhler, 'Für eine Geschichte der Orte und Räume im globalen Zeitalter', in Schröder and Höhler, *Welt-Räume*, pp. 303–13. Unisphere was built to last: a city landmark, it still dazzles visitors in Flushing Meadows-Corona Park in Queens, New York City.

16. 'A Century of Progress' was the theme of the Chicago World Fair in 1933. The New York World's Fair in 1939 and 1940 was similarly optimistic; it was structured around the single idea of 'Building a World of Tomorrow'; see F. T. Kihlstedt, 'Utopia Real-

ized: The World's Fairs of the 1930s', in J. J. Corn (ed.), *Imagining Tomorrow: History, Technology, and the American Future* (Cambridge, MA, and London: MIT Press, 1986), pp. 97–118. The New York World's Fair of 1964/5 conformed to the 1939/40 Fair site; moreover, it demonstrated symbolic power by building the Unisphere on the same foundations that had supported the 'Perisphere' in 1939, a gigantic sphere which housed the exhibition 'The World of Tomorrow'. Together with the 'Trilon', a 700-foot spire, the Perisphere had formed the 1939 Fair's visual logo. Both structures were subsequently dismantled.

17. See, for example, J. Johnson-Freese, *Space as a Strategic Asset* (New York: Columbia University Press, 2007).

18. S. Conrad and S. Randeria (eds), *Jenseits des Eurozentrismus: Postkoloniale Perspektiven in den Geschichts- und Kulturwissenschaften* (Frankfurt: Campus, 2002), p. 12.

19. See, for example, A. Escobar, *Encountering Development: The Making and Unmaking of the Third World* (Princeton, NJ: Princeton University Press, 1995); W. Sachs, *Planet Dialectics: Explorations in Environment and Development* (London: Zed Books, 1999).

20. P. Galison, *Big Science: The Growth of Large Scale Research* (Stanford, CA: Stanford University Press, 1992).

21. *New York World's Fair 1964/1965: Official Guide*, advertisement *U. S. Steel*, p. 179.

22. Moses, 'The Fashioning of a Fair', p. 19. J. Verne, *Around the World in Eighty Days* (New York: Penguin Books and Signet Classics, 2005) [orig. French, *Le tour du monde en quatre-vingts jours* (Paris, 1873)].

23. Moses, 'The Fashioning of a Fair', p. 19.

24. S. Jasanoff, 'Image and Imagination: The Formation of Global Environmental Consciousness', in C. A. Miller and P. N. Edwards (eds), *Changing the Atmosphere: Expert Knowledge and Environmental Governance* (Cambridge, MA: MIT Press, 2001), pp. 309–37; S. Jasanoff, 'Heaven and Earth: The Politics of Environmental Images', in S. Jasanoff and M. L. Martello (eds), *Earthly Politics: Local and Global in Environmental Governance* (Cambridge, MA, and London: MIT Press, 2004), pp. 31–52.

25. J. E. Lovelock, 'The Gaia Hypothesis', in P. Bunyard (ed.), *Gaia in Action: Science of the Living Earth* (Edinburgh: Floris, 1996), pp. 15–33, p. 16 on the composition of the earth's atmosphere that an observer from space would perceive. See also J. E. Lovelock, *Gaia: A New Look at Life on Earth* (Oxford: Oxford University Press, 1979).

26. H. J. Schellnhuber, '"Earth System" Analysis and the Second Copernican Revolution', *Nature*, 402 (1999), pp. C19–C23.

27. R. V. Daniels, *The Fourth Revolution: Transformations in American Society from the Sixties to the Present* (London: Taylor & Francis, 2006). Daniels focuses primarily on social movements; the 'Environmental Revolution' forms no explicit part of his analysis.

28. C. S. Maier, 'Two Sorts of Crisis? The "Long" 1970s in the West and the East', in H. G. Hockerts (ed.), *Koordinaten deutscher Geschichte in der Epoche des Ost-West-Konflikts* (München: R. Oldenbourg Verlag, 2004), pp. 49–62. The past decade saw an explosion of literature, conferences and exhibitions on 1968, social rebellion, student, women and working-class activism. See, for instance, C. Fink, P. Gassert, D. Junker and D. S. Mattern, *1968: The World Transformed* (Cambridge: Cambridge University Press, 1999); G.-R. Horn, *The Spirit of '68: Rebellion in Western Europe and North America, 1956–1976* (Oxford: Oxford University Press, 2007).

29. On the 'threshold effect' see Holdren and Ehrlich, 'Human Population and the Global Environment', p. 290; for examples of environmental historiography arguing in this vein see K. F. Hünemörder, '1972 – Epochenschwelle der Umweltgeschichte?', in F.-J.

Brüggemeier and J. I. Engels (eds), *Natur- und Umweltschutz nach 1945: Konzepte, Konflikte, Kompetenzen* (Frankfurt: Campus, 2005), pp. 124–44.

30. The 1960s have been considered the 'time of awakening', identifying environmental problems and forming the new environmental movement. 'Environmentalism' took its beginning with Rachel Carson's *Silent Spring* (1962), Barry Commoner's *Science & Survival* (1966) and *The Closing Circle* (1971), and Paul Ehrlich's *The Population Bomb* (1968), replacing traditional conservationist movements; see for the US, M. H. Lytle, *The Gentle Subversive: Rachel Carson, Silent Spring, and the Rise of the Environmental Movement* (New York: Oxford University Press, 2007); P. C. Murphy, *What a Book Can Do: The Publication and Reception of Silent Spring* (Amherst and Boston, MA: University of Massachusetts Press, 2005); M. Egan, *Barry Commoner and the Science of Survival: The Remaking of American Environmentalism* (Cambridge, MA: MIT Press, 2007); T. Robertson, *The Malthusian Moment: Global Population Growth and the Birth of American Environmentalism* (New Brunswick, NJ: Rutgers University Press, 2012). The year 1970 marks the beginning of the 'environmental era' in the Western world. Concerning the 'quality-of-life' movement as an American middle- and upper-class phenomenon see F. Buell, *From Apocalypse to Way of Life: Environmental Crisis in the American Century* (New York and London: Routledge, 2003); S. P. Hays, *Beauty, Health and Permanence: Environmental Politics in the United States, 1955–1985* (Cambridge: Cambridge University Press, 1987); H. K. Rothman, *Saving the Planet: The American Response to the Environment in the Twentieth Century* (Chicago, IL: Ivan R. Dee, 2000).

31. F. Turner, *From Counterculture to Cyberculture: Stewart Brand, the Whole Earth Network, and the Rise of Digital Utopianism* (Chicago, IL: University of Chicago Press, 2006). The survivalist movement was based on the contention that the earth's stock of resources was limited. Survivalism prescribed drastic, multidimensional action to prevent global disaster. See J. S. Dryzek, *The Politics of the Earth: Environmental Discourses* (Oxford: Oxford University Press, 1997).

32. The 'Prometheans' or 'Cargoists' expressed faith in the progress of science and technology to maintain affluence. The term Cargoism originates from ethnological studies of the 'Cargo Cults' of Melanesian islanders in the South Pacific and their traditional beliefs in societal regeneration through a mythic charismatic leader or god delivering cargo (shiploads of goods) from the West. On 'Space Age Cargo Cults' see W. R. Catton, *Overshoot: The Ecological Basis of Revolutionary Change* (Urbana, Chicago, IL, and London: University of Illinois Press, 1980), pp. 187–95. On the pessimism and technocratic optimism of the era see E. Mendelsohn, 'The Politics of Pessimism: Science and Technology Circa 1968', in Y. Ezrahi, E. Mendelsohn and H. Segal (eds), *Technology, Pessimism, and Postmodernism* (Dordrecht: Kluwer, 1994), pp. 151–73.

33. On the 'years of decision' see Ehrlich, *The Population Bomb*, Foreword by David Brower, p. 13. On the 'revolution' see M. Nicholson, *The Environmental Revolution: A Guide for the New Masters of the World* (New York: McGraw-Hill, 1970).

34. P. Kupper, 'Die "1970er Diagnose". Grundsätzliche Überlegungen zu einem Wendepunkt der Umweltgeschichte', *Archiv für Sozialgeschichte*, 43 (2003), pp. 325–48.

35. Andrew Jamison speaks of the 'age of ecological innocence' coming to an abrupt end with the oil price shock in 1974; A. Jamison, *The Making of Green Knowledge: Environmental Politics and Cultural Transformation* (Cambridge: Cambridge University Press, 2001), p. 87.

36. *Man and His Future: A Ciba Foundation Volume*, ed. G. Wolstenholme (London: J.

& A. Churchill, 1963); H. Kahn and A. J. Wiener, *The Year 2000: A Framework for Speculation on the Next Thirty-three Years* (New York: Macmillan, 1967); H. Kahn, W. Brown and L. Martel, *The Next 200 Years: A Scenario for America and the World* (New York: Morrow, 1976). On the proliferation of future studies see P. Moll, *From Scarcity to Sustainability: Futures Studies and the Environment. The Role of the Club of Rome* (Frankfurt: Peter Lang Verlag, 1991); A. Schmidt-Gernig, 'The Cybernetic Society: Western Future Studies of the 1960s and 1970s and their Predictions for the Year 2000', in R. Layard and R. N. Cooper (eds), *What the Future Holds: Insights from Social Science* (2002; Cambridge, MA: MIT Press, 2003), pp. 233–59.

37. 'New Age' environmentalism regarded all life as cooperative instead of competitive. Highly contested were the teleology of a planetary homeostasis implicit in this concept as well as the insistence on altruism as the principle of organic cooperation, rejecting theories of natural selection. A. Ross, *Strange Weather: Culture, Science and Technology in the Age of Limits* (London: Verso, 1991), ch. 1: 'New Age – A Kinder, Gentler Science?', pp. 15–74.

38. 'Environmental Era', 'Age of Ecology' and also 'Environmental Decade'; see for instance R. N. L. Andrews, *Managing the Environment, Managing Ourselves: A History of American Environmental Policy* (New Haven, CT, and London: Yale University Press, 1999), p. 228. In 1977 Donald Worster dated the beginning of the 'Age of Ecology' back to the ignition of the first atomic bomb in Alamogordo, New Mexico, in July 1945; D. Worster, *Nature's Economy: A History of Ecological Ideas* (Cambridge, MA, and New York: Cambridge University Press, 1977).

39. S. Chaney, *Nature of the Miracle Years: Conservation in West Germany, 1945–1975* (New York and Oxford: Berghahn Books, 2008); J. Lachmund, *Greening Berlin: The Co-Production of Science, Politics and Urban Nature* (Cambridge, MA, and London: MIT Press, 2013).

40. W. Cronon (ed.), *Uncommon Ground: Rethinking the Human Place in Nature* (1995; New York and London: W. W. Norton, 1996).

41. W. Cronon, 'The Trouble with Wilderness or, Getting Back to the Wrong Nature', *Environmental History*, 1 (1996), pp. 7–28; J. B. Callicott and M. P. Nelson (eds), *The Great New Wilderness Debate: An Expansive Collection of Writings Defining Wilderness from John Muir to Gary Snyder* (Athens, GA: University of Georgia Press, 1998); P. Kupper, *Creating Wilderness: A Transnational History of the Swiss National Park* (New York and Oxford: Berghahn Books, 2014); M. Lewis (ed.), *American Wilderness: A New History* (Oxford and New York: Oxford University Press, 2007). On the relation of environmentalism and religion in the motif of the paradisiacal garden see C. Merchant, *Reinventing Eden: The Fate of Nature in Western Culture* (New York and London: Routledge, 2004); T. R. Dunlap, *Faith in Nature: Environmentalism as Religious Quest* (Seattle, WA, and London: University of Washington Press, 2004).

42. W. Beinart and L. Hughes, *Environment and Empire* (Oxford and New York: Oxford University Press, 2007); P. Coates, *Nature: Western Attitudes since Ancient Times* (Cambridge: Polity Press, 1998); D. J. Hughes, *An Environmental History of the World: Humankind's Changing Role in the Community of Life* (London and New York: Routledge, 2001); J. R. McNeill, *Something New under the Sun: An Environmental History of the Twentieth Century* (London: Penguin Books, 2001); J. Radkau, *Nature and Power: A Global History of the Environment* (Cambridge and New York: Cambridge University Press, 2008) [orig. German *Natur und Macht: Eine Weltgeschichte der Umwelt* (München: C. H. Beck, 2002)].

43. D. J. Haraway, *Modest_Witness@Second_Millenium.FemaleMan©_Meets_OncoMouse™: Feminism and Technoscience* (New York and London: Routledge, 1997), p. 99.

44. At the turn of the twentieth century, nature was protected in the form of striking singular 'monuments' or spectacular or romantic sceneries, sacred, spiritually uplifting and enlightening. The National Park is often used to refer to the 'Fortress Model' of conservation, an idealized version of a 'wilderness' fenced off, a nationally managed nature preserve, from which indigenous people were evicted, thus fitting into the broader historical context of the settlement of the American West. Meanings of nature and notions of nature protection and conservation were not fixed but just shaped at the time, as was the role of the nation in nature protection, fostering ideas of civilization, patriotic self-assurance, progressivism and exceptionalism. B. Gissibl, S. Höhler and P. Kupper (eds), *Civilizing Nature: National Parks in Global Historical Perspective* (New York and Oxford: Berghahn Books, 2012).

45. M. Lewis, *Inventing Global Ecology* (Athens, OH: Ohio University Press, 2004).

46. P. R. Ehrlich, *The Population Bomb* (1968; New York: Ballantine, 1969), p. 65.

47. Ehrlich, *The Population Bomb*, p. 66.

48. Ehrlich, *The Population Bomb*, p. 66.

49. M. Murphy, *Sick Building Syndrome and the Problem of Uncertainty: Environmental Politics, Technoscience, and Women Workers* (Durham, NC: Duke University Press, 2006); M. Farish, *The Contours of America's Cold War* (Minneapolis, MN, and London: University of Minnesota Press, 2010).

50. P. Anker, *From Bauhaus to Ecohouse: A History of Ecological Design* (Baton Rouge, LA: Louisiana State University Press, 2010); M. Hård and A. Jamison, *Hubris and Hybrids: A Cultural History of Technology and Science* (New York and London: Routledge, 2005); F. A. Jørgensen, *Making a Green Machine: The Infrastructure of Beverage Container Recycling* (New Brunswick, NJ: Rutgers University Press, 2011); G. Mitman, M. Murphy and C. Sellers (eds), *Landscapes of Exposure: Knowledge and Illness in Modern Environments* (*Osiris*, 19 (2004)).

51. On Cold War technosciences and their environments see *Earth Sciences in the Cold War* (Special Guest-Edited Issue, *Social Studies of Science*, 33 (2003)), pp. 629–819; J. D. Hamblin, *Arming Mother Nature: The Birth of Catastrophic Environmentalism* (Oxford and New York: Oxford University Press, 2013); J. R. McNeill and C. Unger (eds), *Environmental Histories of the Cold War* (New York: Cambridge University Press, 2010); J. D. Hamblin, *Oceanographers and the Cold War: Disciples of Marine Science* (Seattle, WA: University of Washington Press, 2005); S. Kirsch, *Proving Grounds: Project Plowshare and the Unrealized Dream of Nuclear Earthmoving* (New Brunswick, NJ: Rutgers University Press, 2005); E. Russell, *War and Nature: Fighting Humans and Insects with Chemicals from World War I to Silent Spring* (New York: Cambridge University Press, 2001); R. P. Tucker and E. Russell (eds), *Natural Enemy, Natural Ally: Toward an Environmental History of War* (Corvallis, OR: Oregon State University Press, 2004).

52. B. Latour, *Politics of Nature: How to Bring the Sciences into Democracy* (Cambridge, MA, and London: Harvard University Press, 2004) [orig. French *Politiques de la nature: comment faire entrer les sciences en démocratie*, Paris 1999], p. 25. On the 'modern constitution' see B. Latour, *We Have Never Been Modern* (Cambridge, MA: Harvard University Press, 1993) [orig. French *Nous n'avons jamais été modernes: Essais d'anthropologie symétrique*, Paris 1991], pp. 13 ff.

53. Latour, *Politics of Nature*, p. 18.

54. Latour, *Politics of Nature*, pp. 18, 20.

55. Bruno Latour forged the term 'technoscience' from 'science and technology' in his earlier work on *Science in Action* to describe the collapsing boundaries between science and society and to include all the messy social elements of scientific activity. B. Latour, *Science in Action: How to Follow Scientists and Engineers though Society* (Cambridge, MA: Harvard University Press, 1987), pp. 29, 174–5.

56. Haraway, *Modest_Witness*, p. 137.

57. B. Ward and R. Dubos, *Only One Earth: The Care and Maintenance of a Small Planet. An Unofficial Report Commissioned by the Secretary-General of the United Nations Conference on the Human Environment, Prepared with the Assistance of a 152-Member Committee of Corresponding Consultants in 58 Countries* (New York: Norton, 1972), pp. xvii–xviii.

58. B. Ward, *Spaceship Earth* (New York: Columbia University Press, 1966).

59. R. Buckminster Fuller, *Operating Manual for Spaceship Earth* (1969; New York: E. P. Dutton & Co., 1971). See J. Krause, 'Buckminster Fullers Vorschule der Synergetik', in R. Buckminster Fuller, *Bedienungsanleitung für das Raumschiff Erde und andere Schriften*, ed. J. Krausse (Amsterdam and Dresden, 1998), pp. 213–306, on p. 252, and P. Anker, 'Buckminster Fuller as Captain of Spaceship Earth', *Minerva*, 45 (2007), pp. 417–34 for Fuller's earlier use of the term.

60. K. E. Boulding, 'The Economics of the Coming Spaceship Earth', in H. Jarrett (ed.), *Environmental Quality in a Growing Economy*, Essays from the Sixth RFF Forum on Environmental Quality held in Washington 8 and 9 March 1966 (Baltimore, MD: Johns Hopkins University Press, 1966), pp. 3–14, on p. 3.

61. Boulding, 'The Economics of the Coming Spaceship Earth', pp. 4, 9.

62. In 1960 the concept of 'cyborg' or cybernetic organism as a hybrid of machine and organism had been proposed for space travel, imagining the organic body as an incorporation or extension of technology, and vice versa, to adapt human beings to new environments. M. E. Clynes and N. S. Kline, 'Cyborgs and Space', *Astronautics* (September 1960), pp. 26–7, 74–6. On the cyborg not as fatal hybridization but as a promising concept in feminist technoscience studies, a term appropriated to come to grips with the fractured identities of bodies and machines and the blurring of boundaries between social realities and fictions, see D. J. Haraway, 'A Cyborg Manifesto: Science, Technology, and Socialist-Feminism in the Late Twentieth Century', in D. J. Haraway, *Simians, Cyborgs, and Women: The Reinvention of Nature* (London: Free Association Books, 1991), pp. 149–81.

63. P. R. Ehrlich and R. L. Harriman, *How to Be a Survivor: A Plan to Save Spaceship Earth* (New York: Ballantine, 1971).

64. Foucault, 'Of Other Spaces', p. 23.

65. M. Foucault, *Security, Territory, Population: Lectures at the College de France 1977–1978* (New York: Picador and Palgrave Macmillan, 2009); M. Foucault, *The Birth of Biopolitics: Lectures at the College de France, 1978–1979* (Basingstoke: Palgrave Macmillan, 2008).

66. Following Buckminster Fuller's publication of *Operating Manual for Spaceship Earth* in 1969 Spaceship Earth was translated into European languages, for instance into the German 'Raumschiff Erde', the Dutch 'ruimteschip aarde' and the French 'astronef terre'. See H. J. Achterhuis, 'Van Moeder Aarde tot ruimteschip: humanisme en milieucrisis', in H. J. Achterhuis, *Natuur tussen mythe en techniek* (Baarn: Ambo, 1995), pp. 41–64; Y. Friedman, *Utopies réalisables*, new edn (Paris: Éditions de l'éclat, 2000), ch.

9: 'La ville globale', pp. 185–6. I have not looked at Eastern European, Soviet and Asian contexts.

67. P. R. Ehrlich, A. H. Ehrlich and J. P. Holdren, *Ecoscience: Population, Resources, Environment* (1970; San Francisco, CA: W. H. Freeman, 1977).

68. With regard to transitions from a nurturing earth to an earth system I will not consider gender in any methodical way. I take up gender as a structural category in all chapters to address different gender aspects of Spaceship Earth: hierarchical order on board ships (Chapter 2); the dissecting approach to a planetary machinery (Chapter 3); the reproductive nucleus and simple reproductive schemes in population ecology (Chapter 4); homeostasis and balance (Chapter 5).

2 Containment: The Ship as a Figure of Enclosure and Expansion

1. *Silent Running* (USA: Universal, 1972), directed by Douglas Trumbull, starring Bruce Dern in the role of Freeman Lowell.

2. S. Höhler, '"Spaceship Earth": Envisioning Human Habitats in the Environmental Age', *Bulletin of the German Historical Institute*, 42 (2008), pp. 65–85.

3. Ward, *Spaceship Earth*, p. 15.

4. Foucault, 'Of Other Spaces', p. 27.

5. S. Brant, *The Ship of Fools* (New York: Dover, 1962) [orig. German *Das Narrenschiff*, Basel, 1494].

6. H. Blumenberg, *Schiffbruch mit Zuschauer: Paradigma einer Daseinsmetapher* (1979; Frankfurt: Suhrkamp, 1997).

7. Blumenberg, *Schiffbruch mit Zuschauer*, p. 45 (translations are mine). See for instance the history and stories of the shipwreck of the French frigate *Méduse* off the Atlantic coast of Africa in 1816 and of the subsequent two weeks of horror and resignation of almost a hundred and fifty survivors thronged on a fragile raft involving the sacrileges of mutiny and cannibalism. Théodore Géricault depicted the situation on canvas in 1819 ('Le Radeau de la Méduse'). K. Heinrich, 'Das Floß der Medusa', in R. Schlesier (ed.), *Faszination des Mythos: Studien zu antiken und modernen Interpretationen* (Basel and Frankfurt: Stroemfeld and Roter Stern, 1991), pp. 335–98.

8. Blumenberg, *Schiffbruch mit Zuschauer*, p. 23.

9. Blumenberg, *Schiffbruch mit Zuschauer*, p. 45. In the nineteenth and twentieth century the nautical journey involved less fortune than control, although Blumenberg acknowledges that quantitatively, the nineteenth century had been the epoch of shipwrecks. Ibid., p. 72.

10. R. Barthes, 'The *Nautilus* and the Drunken Boat', in R. Barthes, *Mythologies* (1957; New York: Hill and Wang, 1972), pp. 65–7.

11. J. Verne, *20,000 Leagues Under the Sea* (New York: Penguin Books and Signet Classics, 2001) [orig. French *Vingt Mille Lieues Sous Les Mers*, Paris, 1870], pp. 79–86.

12. 'The great obsession of the nineteenth century was, as we know, history.' Foucault, 'Of Other Spaces', p. 22. On miniatures see G. Bachelard, *The Poetics of Space: The Classic Look at How We Experience Intimate Places* (Boston, MA: Beacon Press, 1994) [orig. French *La poétique de l'espace*, Paris 1958]. Literary scholar Thomas Richards elaborated on how nineteenth-century activities were geared towards conservation through libraries, museums and archives, in a veritable 'obsession with gathering and ordering information' against the chaos and confusion of time, or rather, against increasing entropy. According to Richards, the *Nautilus* was an ordering device, an archive. T.

Richards, *The Imperial Archive: Knowledge and the Fantasy of Empire* (London and New York: Verso, 1993), on p. 9; on the *Nautilus* see pp. 115–23.

13. P. Sloterdijk, *Sphären*, Vol. 2: *Globen (Makrosphärologie)* (Frankfurt: Suhrkamp, 1999), ch. 3: 'Archen. Zur Ontologie des ummauerten Raumes', pp. 251 ff. (translation is mine).

14. Sloterdijk, *Sphären*, p. 251 f.

15. A. Schmidt, *Die Gelehrtenrepublik: Kurzroman aus den Roßbreiten* (1957; Frankfurt: Fischer, 2004).

16. Schmidt, *Die Gelehrtenrepublik*, p. 98 (translations are mine). Three-miles long and one-and-a-half-miles wide, the IRAS combines a large ship and an artificial oval island. Roughly 1,000 artists, poets, painters and scientists live on board at any one time, supported by stipends.

17. Schmidt, *Die Gelehrtenrepublik*, pp. 121, 122. The motif of the island is closely related to the ship with regard to enclosure and autarky. The island figures prominently not only in literature and cultural studies but also in studies of ecological systems behaviour as a theoretical and also practical model. While island narratives merit further exploration as to the stories of autonomy and isolation they tell I concentrate on ship narratives apart from the example given here, the Noah/Robinson theme in Chapter 3 and the ecological 'island' of Biosphere 2 in Chapter 5.

18. Blumenberg, *Schiffbruch mit Zuschauer*, p. 78 ff.

19. L. Nayder, 'Sailing Ships and Steamers, Angels and Whores: History and Gender in Joseph Conrad's Maritime Fiction', in M. S. Creighton and L. Norling (eds), *Iron Men, Wooden Women: Gender and Seafaring in the Atlantic World, 1700–1920* (Baltimore, MD, and London: Johns Hopkins University Press, 1996), pp. 189–203, on pp. 200, 191.

20. H. M. Rozwadowski, 'Small World: Forging a Scientific Maritime Culture for Oceanography', *Isis*, 87 (1996), pp. 409–29, particularly on p. 410.

21. M. S. Creighton, 'Davy Jones's Locker Room: Gender and the American Whaleman, 1830–1870', in Creighton and Norling, *Iron Men, Wooden Women*, pp. 118–37, on pp. 125, 121. On the relations of space and gender and on spatial order and distributions of power along gender lines see D. Massey, *Space, Place, and Gender* (Cambridge: Polity Press, 1994).

22. Nayder, 'Sailing Ships', p. 191.

23. Nayder, 'Sailing Ships', p. 201.

24. Nayder, 'Sailing Ships', p. 197. An extreme all-enclosed, all-male community was portrayed in the German movie *Das Boot*, which features the claustrophobic situation on the German U-boat *U96* navigating the terror of World War II. *Das Boot* (Germany: Bavaria Atelier/Radiant Film/WDR, 1981), directed by Wolfgang Petersen, starring Jürgen Prochnow, Herbert Grönemeyer and Klaus Wennemann.

25. See Blumenberg, *Schiffbruch mit Zuschauer*, p. 9 ff., on seafaring as a violation of borders and the sea as the sphere of irregularity and anarchy.

26. H. Melville, *Moby-Dick, or, The Whale* (1851; New York and London: Penguin Books, 1992); M. Twain, *The Adventures of Huckleberry Finn*, adapted by Matthew Francis (1884; London and New York: S. French, 1998); E. Hemingway, *The Old Man and the Sea* (1952; New York: Scribner, 1996).

27. B. Traven, *The Death Ship: The Story of an American Sailor* (London: Chatto & Windus, 1934) [orig. German *Das Totenschiff*, 1926].

28. See S. A. Ogilvie and S. Miller, *Refuge Denied: The St. Louis Passengers and the Holo-*

caust (Madison, WI: University of Wisconsin Press, 2006); G. Thomas and M. Morgan Witts, *Voyage of the Damned* (New York: Stein and Day, 1974). A movie on the subject, titled *Voyage of the Damned*, was directed by Steward Rosenberg (USA: Avco, 1976).

29. Y. Martel, *Life of Pi* (New York: Harcourt, 2001).

30. Martel, *Life of Pi*, pp. 120, 223, 217.

31. Based on a story by John Steinbeck, *The Lifeboat* (USA: 20th Century Fox, 1944), directed by Alfred Hitchcock, starring Tallulah Bankhead and William Bendix, pictures a lifeboat situation of cooperation and competition for survival in the midst of the Second World War.

32. Martel, *Life of Pi*, p. 164.

33. J. Barnes, *A History of the World in 10½ Chapters* (1989; New York: Vintage, 1990), ch. 1: 'The Stowaway', pp. 1–30.

34. Barnes, *A History of the World in 10½ Chapters*, pp. 4, 7, 10.

35. Webb, *The Great Frontier*, p. 1.

36. Webb, *The Great Frontier*, pp. 8–9.

37. Webb, *The Great Frontier*, p. 9.

38. S. Harding, *Is Science Multicultural? Postcolonialisms, Feminisms, and Epistemologies* (Bloomington, IN: Indiana University Press, 1998), p. 23 ff., on the 'Miracle' see p. 26.

39. Haraway, *Modest_Witness*, pp. 135–41.

40. Harding, *Is Science Multicultural?*, p. 23.

41. Harding, *Is Science Multicultural?*, pp. 146 ff.; D. J. Haraway, 'Situated Knowledges: The Science Question in Feminism and the Privilege of Partial Perspective', in D. J. Haraway, *Simians, Cyborgs, and Women: The Reinvention of Nature* (London: Free Association Books, 1991), pp. 183–201.

42. E. Mancke, 'Early Modern Expansion and the Politicization of Oceanic Space', *Geographical Review*, 89 (1999), pp. 225–36, on p. 225.

43. Harding, *Is Science Multicultural?*, p. 39, particularly ch. 3: 'Voyages of Discovery: Imperial and Scientific', pp. 39 ff.

44. See, for example, R. Porter and M. Teich (eds), *The Scientific Revolution in National Context* (Cambridge and New York: Cambridge University Press, 1992). For a critical perspective see S. Shapin, *The Scientific Revolution* (Chicago, IL: University of Chicago Press, 1996).

45. W. Eisler, *The Furthest Shore: Images of Terra Australis from the Middle Ages to Captain Cook* (Cambridge: Cambridge University Press, 1995); G. A. Mawer, *South by Northwest: The Magnetic Crusade and the Contest for Antarctica* (Edinburgh: Birlinn, 2006); G. Williams, *Voyages of Delusion: The Quest for the Northwest Passage* (New Haven, CT: Yale University Press, 2003).

46. For a survey of historical and sociological theories of space see S. Rau, *Räume: Konzepte, Wahrnehmungen, Nutzungen* (Frankfurt: Campus, 2013); *Raum: Ein interdisziplinäres Handbuch*, ed. S. Günzel (Stuttgart and Weimar: Verlag J. B. Metzler, 2010).

47. M.-N. Bourguet, C. Licoppe and H. O. Sibum, *Instruments, Travel and Science: Itineraries of Precision from the Seventeenth to the Twentieth Century* (London: Routledge, 2002); N. Jardine, J. A. Secord and E. C. Spary (eds), *Cultures of Natural History* (Cambridge: Cambridge University Press, 1996); M. L. Pratt, *Imperial Eyes: Travel Writing and Transculturation* (London and New York: Routledge, 1992).

48. B. Latour, 'Centres of Calculation', in Latour, *Science in Action*, pp. 215–57.

49. Harding, *Is Science Multicultural?*, pp. 50, 35.

50. D. R. Headrick, *The Tentacles of Progress: Technology Transfer in the Age of Imperialism, 1850–1940* (New York and Oxford: Oxford University Press, 1988); D. K. Lahiri-Choudhury, *Telegraphic Imperialism: Crisis and Panic in the Indian Empire, c. 1830–1920* (Basingstoke: Palgrave Macmillan, 2010).

51. Richards, *The Imperial Archive*, p. 1.

52. Richards, *The Imperial Archive*, p. 4.

53. Richards, *The Imperial Archive*, p. 6. See also J. C. Scott, *Seeing Like a State: How Certain Schemes to Improve the Human Condition Have Failed* (New Haven, CT: Yale University Press, 1998).

54. R. Sorrenson, 'The Ship as a Scientific Instrument in the Eighteenth Century', in H. Kuklick and R. E. Kohler (eds), *Science in the Field* (*Osiris*, 11 (1996)), pp. 221–36, on p. 229. On scientific representations of the deep sea in nineteenth- and twentieth-century oceanography see S. Höhler, 'Depth Records and Ocean Volumes: Ocean Profiling by Sounding Technology, 1850–1930', *History and Technology*, 18 (2002), pp. 119–54. Helen Rozwadowski brought forward a similar argument concerning technology's mediating role in shaping the image of the mid-nineteenth-century deep-sea floor. Exploring deep-sea research in the context of the Atlantic telegraph cable projects in the 1850s and 1860s she argues that the 'ocean-scape' took shape only within a complex interaction of instruments, methods, interpretations and motivations for depth measurement; H. M. Rozwadowski, *Fathoming the Ocean: The Discovery and Exploration of the Deep Sea* (Cambridge, MA, and London: Belknap Press of Harvard University Press, 2005). On techniques of 'accumulating time and space' by way of 'immutable mobiles' or stable and transportable charts, see B. Latour, 'Drawing Things Together', in M. Lynch and S. Woolgar (eds), *Representation in Scientific Practice* (Cambridge, MA, and London: MIT Press, 1990), pp. 19–68, on pp. 31 ff. Arguing that stability in transport and translation of an abstract concept matters more than its unity and integrity Latour also elaborated the notion of 'optical consistency'; see B. Latour, 'Visualization and Cognition: Thinking with Eyes and Hands', *Knowledge and Society: Studies in the Sociology of Culture Past and Present*, 6 (1986), pp. 1–40, particularly pp. 7–9.

55. Richards, *The Imperial Archive*, p. 113.

56. On Humboldtian Science see S. F. Cannon, *Science in Culture: The Early Victorian Period* (New York: Dawson, 1978); M. Dettelbach, 'Humboldtian Science', in N. Jardine, J. A. Secord and E. C. Spary (eds), *Cultures of Natural History* (Cambridge: Cambridge University Press, 1996), pp. 287–304; for 'atmospheric physics' based on the balloon or 'airship': S. Höhler, *Luftfahrtforschung und Luftfahrtmythos: Wissenschaftliche Ballonfahrt in Deutschland, 1880–1910* (Frankfurt: Campus, 2001). *Naus*, the Greek term for ship, linked aeronautics etymologically to navigation at sea.

57. Verne, *20,000 Leagues Under the Sea*, pp. 303–5.

58. M. F. Maury, *The Physical Geography of the Sea* (1855; New York and London: Harper and Sampson Low, 1859); P. H. Kylstra and A. Meerburg, 'Jules Verne, Maury and the Ocean', in *Challenger Expedition Centenary,* Proceedings of the Second International Congress on the History of Oceanography, Edinburgh, September 12–20, 1972 (Edinburgh: Royal Society, 1972), vol. 1, pp. 243–51.

59. Verne, *20,000 Leagues Under the Sea*, p. 284.

60. Verne, *20,000 Leagues Under the Sea*, pp. 87–8.

61. Verne, *20,000 Leagues Under the Sea*, p. 99.

62. Verne, *20,000 Leagues Under the Sea*, p. 99.

63. P. N. Limerick, *The Legacy of Conquest: The Unbroken Past of the American West* (New

York and London: W. W. Norton, 1987), pp. 322–3.

64. F. J. Turner, *The Frontier in American History* (New York: Henry Holt & Co., 1921), ch. 1: 'The Significance of the Frontier in American History', p. 1; W. H. Goetzmann and W. N. Goetzmann, *The West of the Imagination* (New York and London: W. W. Norton, 1986); W. Stegner, *The American West as Living Space* (Ann Arbor, MI: University of Michigan Press, 1987). A mythological and symbolic interpretation of the American West was first given by Henry Nash Smith in his founding text *Virgin Land: The American West as Symbol and Myth* (1950; New York: Vintage Books, 1957). Patricia Limerick followed the question of fact and (literary) fiction, of myth and its symbols: P. N. Limerick, 'Making the Most of Words: Verbal Activity and Western America', in W. Cronon, G. Miles and J. Gitlin (eds), *Under an Open Sky: Rethinking America's Western Past* (New York and London: W. W. Norton, 1992), pp. 167–84, particularly pp. 170 ff.

65. W. Cronon, G. Miles and J. Gitlin, 'Becoming West: Toward a New Meaning for Western History', in Cronon et al., *Under an Open Sky*, pp. 3–27, on p. 10; J. Gitlin, 'On the Boundaries of Empire', in Cronon et al., *Under an Open Sky*, pp. 71–89. Gitlin considers Turner's two-dimensional model of the frontier as a line of settlement as flattening the view of a whole array of connections. He pleads for a zone model of various distinct frontier zones, where the traditional frontier is something like a 'graft' or overlay on the history of some regions (p. 88).

66. Webb, *The Great Frontier*, p. 2.

67. Webb, *The Great Frontier*, pp. 2–3 (original emphasis). Webb distinguishes the 'frontier' as used in the US from the 'boundary' as for instance the line separating the US from Canada, and from the 'border' as for example the heavily guarded border separating the US from Mexico. See also Webb's *Divided We Stand: The Crisis of a Frontierless Democracy* (New York: Farrar & Rinehart, 1937).

68. Cronon et al. 'Becoming West', p. 6 (original emphasis). Compare Limerick, *Legacy*, p. 26, and Limerick, 'Making the Most of Words', p. 171 on repetition.

69. In accordance with Cronon et al., stating: 'The West of the popular imagination, unlike the West of the scholars, is an almost timeless sort of place'. Cronon et al., 'Becoming West', p. 4.

70. Limerick, 'Making the Most of Words', p. 171.

71. G. H. Nobles, 'Straight Lines and Stability: Mapping the Political Order of the Anglo-American Frontier', *Journal of American History*, 80 (1993), pp. 9–35.

72. On myths and metaphors of the 'new frontier' see C. Abbott, *Frontiers Past and Future: Science Fiction and the American West* (Lawrence, KS: University Press of Kansas, 2006); D. E. Nye, *America as Second Creation. Technology and Narratives of New Beginnings* (Cambridge, MA: MIT Press, 2004); D. M. Wrobel, *The End of American Exceptionalism: Frontier Anxiety from the Old West to the New Deal* (Lawrence, KS: University Press of Kansas, 1993).

73. W. Natter and J. P. Jones, 'Identity, Space, and other Uncertainties', in G. Benko and U. Strohmayer (eds), *Space and Social Theory: Interpreting Modernity and Postmodernity* (Oxford: Basil Blackwell, 1997), pp. 141–61, on p. 147.

74. Louis L'Amour, American author of Western fiction, as quoted in Limerick, *Legacy*, p. 32. L'Amour made this contribution to the collection 'Does America Still Exist?' (*Harper's Magazine*, March 1984).

75. C. Penley, *NASA/Trek: Popular Science and Sex in America* (London and New York: Verso, 1997); G. K. O'Neill, *The High Frontier: Human Colonies in Space* (New York:

Morrow, 1976).

76. See for instance R. Miller, *The Dream Machines: An Illustrated History of the Spaceship in Art, Science and Literature*, Foreword by Arthur C. Clarke (Malabar, FL: Krieger, 1993).

77. Webb, *The Great Frontier*, p. 349.

78. Webb, *The Great Frontier*, p. 350.

79. Osborn, *The Limits of the Earth*, p. 78.

80. W. McDougall, *... the Heavens and the Earth: A Political History of the Space Age* (1985; Baltimore, MD: Johns Hopkins University Press, 1997); A. A. Siddiqi, *The Soviet Space Race with Apollo* (Gainesville, FL: University Press of Florida, 2003); S. J. Dick and R. D. Launius (eds), *Critical Issues in the History of Spaceflight* (Washington, DC: NASA History Division, Office of External Relations, 2006); K. Werth, *Ersatzkrieg im Weltraum: Das US-Raumfahrtprogramm in der Öffentlichkeit der 1960er Jahre* (Frankfurt: Campus, 2006).

81. Johnson-Freese, *Space as a Strategic Asset*.

82. On the role of the new media in forging the self-understanding of Western democracy during the Cold War and also covering the moon landing, with a focus on Germany both East and West, see T. T. Lindenberger (ed.), *Massenmedien im Kalten Krieg: Akteure, Bilder, Resonanzen* (Köln: Böhlau, 2006). On the popular fantasies of space exploration as promoted by NASA propaganda and emerging mass media see G. J. De-Groot, *Dark Side of the Moon: The Magnificent Madness of the American Lunar Quest* (New York: New York University Press, 2006).

83. V. Bush, *Science – The Endless Frontier* (Washington, DC: United States Government Printing Office, 1945); see also G. P. Zachary, *Endless Frontier: Vannevar Bush, Engineer of the American Century* (Cambridge, MA, and London: MIT Press, 1999). On post-war science and research in the US see also D. L. Kleinman, *Politics on the Endless Frontier: Postwar Research Policy in the United States* (Durham, NC: Duke University Press, 1995); Z. Wang, *In Sputnik's Shadow: The President's Science Advisory Committee and Cold War America* (New Brunswick, NJ: Rutgers University Press, 2008); D. H. DeVorkin, *Science with a Vengeance: How the Military Created the US Space Sciences after World War II* (New York and Berlin: Springer, 1992).

84. J. F. Kennedy, Special Message to the Congress on Urgent National Needs, 25 May 1961, John F. Kennedy Presidential Library & Museum, Historical Resources, at http://www.jfklibrary.org/Asset-Viewer/Archives/JFKPOF-034–030.aspx [accessed 28 August 2014]. See H. E. McCurdy, *Space and the American Imagination* (Washington, DC: Smithsonian Institution Press, 1997); J. M. Logsdon, *The Decision to Go to the Moon: Project Apollo and the National Interest*, Cambridge, MA, and London: MIT Press, 1970); R. C. Seamans Jr, *Project Apollo: The Tough Decisions* (Washington, DC: NASA History Division, 2005).

85. J. F. Kennedy, Address at Rice University on the Nation's Space Effort, 12 September 1962, NASA History Division, Key Documents in the History of Space Policy, at http://history.nasa.gov/spdocs.html [accessed 28 August 2014].

86. Ibid., as for the following quotes.

87. *Jewels of Light: The Stained Glass of Washington National Cathedral*, ed. E. R. Crimi and D. Ney (Washington, DC: Washington National Cathedral Guidebooks, 2004), p. 69; *Washington National Cathedral: Guidebook to the Cathedral* (Washington, DC: Washington National Cathedral Guidebooks, 2004), p. 32. Artist Rodney Winfield designed the space window, his only window in the cathedral. The window was dedicated on 21

April 1974, on the fifth anniversary of the Apollo 11 moon landing.

88. *Washington National Cathedral: Guidebook*, p. 32.
89. *Guide to Gargoyles and Other Grotesques*, ed. A.-C. Fallen (Washington, DC: Washington National Cathedral, 2003), p. 107. Darth Vader belongs to the 'North Grotesques'. He is located at the East face of the Northwest Tower, 183 feet above ground. Thirteen-year-old Chris Rader from Nebraska designed the figure of Darth Vader in a Draw-a-Grotesque competition sponsored by *National Geographic World* in 1985.
90. Haraway, *Modest_Witness*. Remarkable in Haraway's work is the thought that to 'be "made" is not to be "made up"'; rather, to be made is 'about contingency and specificity but not epistemological relativism', p. 99.
91. J. Butler, *Bodies that Matter: On the Discursive Limits of 'Sex'* (New York and London: Routledge, 1993).
92. On myth as semantic construction see U. Greiner-Kemptner and R. F. Riesinger (eds), *Neue Mythographien: Gegenwartsmythen in der interdisziplinären Debatte* (Wien, Köln, Weimar: Böhlau, 1995); A. Völker-Rasor and W. Schmale (eds), *MythenMächte – Mythen als Argument* (Berlin: Berlin Verlag Arno Spitz, 1998).
93. R. Barthes, 'Myth Today', in Barthes, *Mythologies*, pp. 109–59.
94. Barthes, 'Myth Today', pp. 120–1.
95. H. Blumenberg, *Arbeit am Mythos* (Frankfurt: Suhrkamp, 1979).
96. Foucault, 'Of Other Spaces', p. 27.
97. Barthes, 'Myth Today', p. 121.
98. Barthes, 'Myth Today', p. 129.
99. Barthes, 'Myth Today', p. 129.

3 Circulation: Ecological Life Support Systems

1. Schellnhuber, '"Earth System" Analysis', p. C19. On the image of Earth as a 'patient' see also W. Sachs (ed.), *Der Planet als Patient: Über die Widersprüche globaler Umweltpolitik* (Basel: Birkhäuser, 1994).
2. Schellnhuber, '"Earth System" Analysis', p. C19.
3. Schellnhuber, '"Earth System" Analysis', p. C19.
4. B. Weber, 'Ubi Caelum Terrae Se Coniungit: Ein altertümlicher Aufriß des Weltgebäudes von Camille Flammarion', *Gutenberg-Jahrbuch* (1975), pp. 381–408.
5. Foucault, 'Of Other Spaces', p. 22.
6. On historiography as a process of mining the strata of time see R. Koselleck, *Zeitschichten: Studien zur Historik* (Frankfurt: Suhrkamp, 2000).
7. On the history of viewing the earth as a whole see U. Bergermann, I. Otto and G. Schabacher (eds), *Das Planetarische: Kultur – Technik – Medien im postglobalen Zeitalter* (München: Wilhelm Fink, 2010); D. Cosgrove, *Apollo's Eye: A Cartographic Genealogy of the Earth in the Western Imagination* (2001; Baltimore, MD, and London: Johns Hopkins University Press, 2003).
8. L. Schiebinger, *Nature's Body: Gender in the Making of Modern Science* (New Brunswick, NJ: Rutgers University Press, 1993). A classic on Enlightenment science and its dissecting relation to nature is C. Merchant, *The Death of Nature: Women, Ecology, and the Scientific Revolution* (San Francisco, CA: Harper & Row, 1980).
9. Charles Goodwin's concept of a 'historically constituted *architecture for perception*' describes a way of seeing that is institutionalized not only through texts but also through 'tools' that shape perception 'though the way in which they construct representations'.

In his study of the construction of oceanic depth in the interplay of laboratory practices aboard a ship, Goodwin shows how instruments and theories need to come together so that one can actually *see* underwater features; C. Goodwin, 'Seeing in Depth', *Social Studies of Science*, 25 (1995), pp. 237–74, on p. 254, 256 (original emphasis).

10. My discussion of 'biosphere' addresses 'living space' primarily as ecological habitat, not 'living space' as political territory that figured prominently in the debates of political economy and political geography of the early twentieth century. However, habitat ecology and population ecology share common threads with the tradition of biogeography and human geography that elevated the concept of living space to a discursive figure of national and global politics. As I will discuss in Chapter 4, the ecological idea of a globally limited living space raised fundamental questions of 'capacity', which took up questions of living space formulated in political geography earlier in the twentieth century. On 'living space' as a territorial concept in Friedrich Ratzel's biogeography and anthropogeography see W. Natter, 'Friedrich Ratzel's Spatial Turn: Identities of Disciplinary Space and its Borders Between the Anthropo- and Political Geography of Germany and the United States', in H. van Houtum, O. Kramsch and W. Zierhofer (eds), *B/ordering Space* (Aldershot: Ashgate, 2005), pp. 171–85; Jureit, *Das Ordnen von Räumen*.

11. Sloterdijk, *Sphären*, pp. 251 ff. (translation is mine).

12. For a natural science perspective on the biosphere see V. Smil, *The Earth's Biosphere: Evolution, Dynamics, and Change* (Cambridge, MA, and London: MIT Press, 2003); V. Smil, *Harvesting the Biosphere: What We Have Taken From Nature* (Cambridge, MA: MIT Press, 2013). On the earth system see J. C. Briden and T. E. Downing (eds), *Managing the Earth: The Linacre Lectures 2001* (Oxford and New York: Oxford University Press, 2002); J. G. Miller and J. L. Miller, 'The Earth as System', *Behavioral Science*, 27 (1982), pp. 303–22.

13. D. Park, *The Grand Contraption: The World as Myth, Number, and Chance* (Princeton, NJ: Princeton University Press, 2005).

14. G. E. Hutchinson, 'The Biosphere', in *The Biosphere. A Scientific American Book* (San Francisco, CA: W. H. Freeman and Company, 1970), pp. 3–11. The article was first published in *Scientific American*, 223 (1970), pp. 45–53, and was reprinted later in the year in the volume *The Biosphere* from which I quote here.

15. Hutchinson, 'The Biosphere', p. 3.

16. Hutchinson, 'The Biosphere', p. 3.

17. E. Sueß, *Die Entstehung der Alpen* (Wien: Braunmüller, 1875), p. 159 (translations are mine).

18. Sueß, *Die Entstehung der Alpen*, pp. 145, 158. J. Grinevald, 'Sketch for a History of the Idea of the Biosphere', in P. Bunyard (ed.), *Gaia in Action: Science of the Living Earth* (Edinburgh: Floris, 1996), pp. 34–53. For a critical perspective W. Sachs, 'Natur als System. Vorläufiges zur Kritik der Ökologie', *Scheidewege: Jahresschrift für skeptisches Denken*, 21 (1991/2), pp. 83–97.

19. E. Sueß, *Das Antlitz der Erde*, 3 vols (Prag, Wien and Leipzig: F. Tempsky and G. Freytag, 1885–1909), vol. 3, 2. half, pt 4, § 27, p. 739.

20. Sueß, *Das Antlitz der Erde*, vol. 3, 2. half, pt. 4, § 27, p. 739.

21. Sueß, *Das Antlitz der Erde*, vol. 3, 2. half, pt. 4, § 27, p. 777.

22. Sueß, *Das Antlitz der Erde*, vol. 3, 2. half, pt. 4, § 27, p. 742.

23. V. I. Vernadsky, *The Biosphere* (New York and Heidelberg: Copernicus and Springer, 1998 [revised and annotated English translation, orig. Russian *Biosfera*, Leningrad:

Nauka, 1926]). The editors attribute the trifling reception and spreading of Vernadsky's work until the 1980s to the trans-European barrier of the Cold War, the Iron Curtain. According to the editors, the pre-wartimes of the 1930s and 1940s prevented the intended translation of *Biosfera* into English. A French translation from 1929 became the master copy for the first abridged English translation in 1986 that was done in the course of the project of 'Biosphere 2' (see Chapter 5 of this book). Ibid., Foreword by Lynn Margulis et al., pp. 15–17.

24. Vernadsky, *The Biosphere*, p. 91.

25. Vernadsky, *The Biosphere*, Foreword by Lynn Margulis et al., p. 15.

26. Vernadsky, *The Biosphere*, Foreword by Lynn Margulis et al., p. 15.

27. Vernadsky, *The Biosphere*, § 68: 'The Biosphere: An Envelope of the Earth', p. 91, Annotation no. 172. The annotation of the passage also supplies an English translation of the corresponding passage in Sueß's text: '"On the surface of continents it is possible to single out a self-contained biosphere" (Sueß, 1875, p. 159).' Compare the translation in P. R. Samson and D. Pitt (eds), *The Biosphere and the Noosphere Reader: Global Environment, Society and Change* (London and New York: Routledge, 1999), p. 23, which speaks of the biosphere as 'self-maintained': 'On the surface of the continents we can distinguish a self-maintained biosphere [eine selbständige Biosphäre]'.

28. Vernadsky, *The Biosphere*, Foreword, p. 18. On Vernadsky's life and work see G. S. Levit, *Biogeochemistry, Biosphere, Noosphere: The Growth of the Theoretical System of Vladimir Ivanovich Vernadsky (1863–1945)* (Berlin: VWB, 2001).

29. Vernadsky, *The Biosphere*, Introduction by Jacques Grinevald, 'The Invisibility of the Vernadskian Revolution', p. 24: 'Both Vernadsky and Teilhard [Pierre Teilhard de Chardin, French theologist, cosmologist and geologist] were cosmic prophets of globalization'.

30. J. Rifkin, *Biosphere Politics: A New Consciousness for a New Century* (New York: Crown, 1991), part five: 'The Coming of the Biospheric Age', pp. 249 ff., on p. 256 f.

31. James Lovelock, creator of the 'Gaia Hypothesis' of the earth as a single living organism, and his colleague Lynn Margulis also perceived the biosphere as a self-organizing system; Lovelock, *Gaia: A New Look*; J. E. Lovelock 'Gaia as Seen Through the Atmosphere', *Atmospheric Environment*, 6 (1972), pp. 579–80; J. E. Lovelock and L. Margulis, 'Atmospheric Homeostasis by and for the Biosphere – the Gaia Hypothesis', *Tellus*, 26 (1974), pp. 2–10. See Chapter 5 of this book on the merging of the holistic and the technological perception of the earth in the Biosphere 2 project. Biosphere 2 is a perfect example of how the language of systems theory and cybernetics pervaded alternative holistic views of the earth from the beginning.

32. Hutchinson, 'The Biosphere', pp. 8, 5.

33. Hutchinson, 'The Biosphere', pp. 7, 11. Hutchinson considered the production of utilizable fossil fuels within this cycle as an accidental and unfortunate imperfection.

34. Sueß, *Das Antlitz der Erde*, vol. 3, p. 740.

35. D. Cosgrove, 'Environmental Thought and Action: Pre-modern and Post-modern', *Transactions of the Institute of British Geographers*, new series, 15 (1990), pp. 344–58; S. P. Hays, 'From Conservation to Environment: Environmental Politics in the United States since World War Two', *Environmental Review*, 6 (1982), pp. 14–41.

36. *The Biosphere*. A *Scientific American* Book (San Francisco, CA: W. H. Freeman and Company, 1970). Foreword by the editors, September 1970, p. vii f.

37. Commoner, *The Closing Circle*.

38. R. Guha, *Environmentalism: A Global History* (New York: Longman, 2000); U. K.

Heise, *Sense of Place and Sense of Planet: The Environmental Imagination of the Global* (Oxford and New York: Oxford University Press, 2008); A. Rome, *The Genius of Earth Day: How a 1970 Teach-In Unexpectedly Made the First Green Generation* (New York: Hill & Wang, 2013).

39. *The Whole Earth Catalog: Access to Tools* borrowed from the genre of the mail order catalogue that had a long tradition in the US. In the nineteenth century, catalogues first supplied settlers on the American frontier. Stewart Brand, the initiator and editor of the *Whole Earth Catalog*, had an access tool in mind that would offer and promote devices and advice adapted to the new frontier of the counterculture. The first edition of 1968 encompassed a mere sixty-four pages, but grew to several hundred pages within years. See A. Kirk, 'Appropriating Technology. The Whole Earth Catalog and Counterculture Environmental Politics', *Environmental History*, 6 (2001), pp. 374–94; Turner, *From Counterculture to Cyberculture*.

40. 'Blue Planet' as seen from space has since then been circulating in popular culture, in fictional and in non-fictional works. See D. Cosgrove, 'Contested Global Visions: *One-World*, *Whole-Earth*, and the Apollo Space Photographs', *Annals of the Association of American Geographers*, 84 (1994), pp. 270–94; Y. Grab, 'The Use and Misuse of the Whole Earth Image', *Whole Earth Review*, 44 (1985), pp. 18–25; T. Ingold, 'Globes and Spheres: The Topology of Environmentalism', in K. Milton (ed), *Environmentalism: The View from Anthropology* (London and New York: Routledge, 1993), pp. 29–40; D. Woodward, 'The Image of the Spherical Earth', *Perspecta. The Yale Architectural Journal*, 25 (1989), pp. 2–15.

    The earth as seen from space also featured prominently in science fiction works. The movie *2001: A Space Odyssey* (USA: Metro-Goldwyn-Mayer, 1968), directed by Stanley Kubrick, which was based on a novel by Arthur C. Clarke, made use of a full earth image. Similarly, the movie *Silent Running* in 1972 featured a shot of 'Blue Planet' as viewed from space freighter *Valley Forge*. The first picture of Full Earth that was taken by a US space mission dates from December 1972. The photograph was taken on the final voyage to the moon by Apollo 17, which terminated the Apollo programme.

41. The American songwriter and musician Gil Scott-Heron covered the song in 1972. It featured on his album *The Revolution Will Not Be Televised* (Flying Dutchman/RCA, 1974); for the lyrics see http://www.gilscottheron.com/lywhitey.html [accessed 28 August 2014]. On the returning motif of the decent, clean and brave young white male US fighter pilot from Mercury to Apollo see the documentary novel by O. Fallaci, *If the Sun Dies* (New York: Atheneum, 1966) [orig. Italian *Se il Sole muore*, Milan 1965].

42. Krausse, 'Buckminster', p. 271.

43. A. MacLeish, 'A Reflection: Riders On Earth Together, Brothers in Eternal Cold', *New York Times*, 25 December 1968.

44. 'The Apollo 8 Christmas Eve Broadcast', National Space Science Data Center, National Aeronautics and Space Administration, at http://nssdc.gsfc.nasa.gov/planetary/lunar/apollo8_xmas.html [accessed 28 August 2014]. See also DeGroot, *Dark Side*, ch. 12: 'Merry Christmas from the Moon', p. 223 ff.

45. Ward, *Spaceship Earth*, p. vii.

46. Ward, *Spaceship Earth*, p. vii.

47. Ward, *Spaceship Earth*, p. 15.

48. Ward, *Spaceship Earth*, p. 15. Ward borrowed her comparison from Richard Buckminster Fuller, 'who, more clearly than most scientists and innovators, has grasped the implications of our revolutionary technology', ibid. Another early spaceship reference

is W. G. Pollard, *Man on a Space Ship: The Meaning of the Twentieth Century Revolution and the Status of Man in the Twenty-first and After*, Foreword by Joseph B. Platt (Claremont, CA: Claremont Colleges, 1967).

49. E. F. Schumacher, *Small Is Beautiful: A Study of Economics as if People Mattered* (London: Blond Briggs, 1973), pp. 17, 11; on 'technology with a human face' see pp. 136 ff.; on 'intermediate technology' pp. 159 ff. Concerning small-scale compared to large-scale technologies see also Richard Feynman's early vision of what would become nanotechnology: R. P. Feynman, 'There's Plenty of Room at the Bottom. An Invitation to Enter a New Field of Physics', *Engineering and Science*, 23 (1960), pp. 22–36.

50. Ward, *Spaceship Earth*, p. 15.

51. Andrew Jamison speaks of the 'age of ecological innocence' coming to an abrupt end with the oil crises in 1973 and 1974; Jamison, *The Making of Green Knowledge*, p. 87.

52. See, for instance, J. R. Fleming, *Fixing the Sky: The Checkered History of Weather and Climate Control* (New York: Columbia UniversityPress, 2010); M. V. Melosi, *Atomic Age America* (Boston, MA: Pearson, 2012).

53. G. R. Taylor, *The Doomsdaybook* (London: Thames & Hudson, 1970; G. R. Taylor, *The Biological Time-Bomb* (London: Thames & Hudson, 1968); A. Toffler, *Future Shock* (New York: Random House, 1970). See also Mendelsohn, 'The Politics of Pessimism'.

54. Ward, *Spaceship Earth*, pp. 13–14.

55. *The Noah* (USA, 1974), directed by Daniel Bourla, starring Robert Strauss.

56. D. Defoe, *Robinson Crusoe* (1719; New York: Barnes & Noble Classics, 2005).

57. Boulding, 'The Economics', pp. 3, 4. Boulding's programmatic text became a classic. It was reprinted in: K. E. Boulding, *Beyond Economics: Essays on Society, Religion, and Ethics* (Ann Arbor, MI: University of Michigan Press, 1970 [1. paperback edition; orig. 1968]), pp. 275–87, and in H. E. Daly (ed.), *Economics, Ecology, Ethics: Essays Toward a Steady-State Economy* (San Francisco, CA: W. H. Freeman and Company, 1980), pp. 253–63. For the translation of Boulding's seminal text into German see S. Höhler and F. Luks (eds), *Beam us up, Boulding! 40 Jahre 'Raumschiff Erde'*, special issue on the 40th anniversary of Kenneth E. Boulding's 'Operating Manual for Spaceship Earth' (1966), *Vereinigung für Ökologische Ökonomie – Beiträge und Berichte*, 7 (2006).

58. Boulding, 'Economics', p. 3 f.

59. Boulding, 'Economics', p. 3.

60. Boulding, 'Economics', p. 9.

61. Boulding, 'Economics', p. 9.

62. *The Noah*, Noah talking to himself, taking the role of a senior officer.

63. R. F. Dasmann, *Planet in Peril? Man and the Biosphere Today* (Harmondsworth: Penguin Books – Unesco, 1972), back cover. In the decade of its creation Spaceship Earth was often spelled in lower case letters, set in quotation marks, or was used with the definite article.

64. Dasmann, *Planet in Peril?* Foreword by Adriano Buzzati-Traverso, pp. 7–9, quote p. 8. Buzzati-Traverso highlights the 'impending dangers' of the 'environmental crisis' and the 'deterioration of the natural environment'.

65. *Use and Conservation of the Biosphere*, Proceedings of the Intergovernmental Conference of Experts on the Scientific Basis for Rational Use and Conservation of the Resources of the Biosphere, Paris, 4–13 September 1968 (Liège: Unesco, 1970), Address by Mr Guy Gresford, Director for Science and Technology of the United Nations Department for Economic and Social Affairs, pp. 251–7.

66. Dasmann, *Planet in Peril?*, ch. 7: 'An International Programme', pp. 127–32, on p. 127.

67. *Use and Conservation of the Biosphere*, pp. 185–7.

68. *Use and Conservation of the Biosphere*, pp. 177–89.

69. *Use and Conservation of the Biosphere*, p. 252.

70. *Use and Conservation of the Biosphere*, p. 252, and foreword: 'the biosphere was taken as that part of the world in which life can exist'.

71. *Use and Conservation of the Biosphere*, p. 210. See also p. 256: 'The biosphere and man's place in it must be envisaged as a whole'.

72. *Use and Conservation of the Biosphere*, pp. 191–235, Final Report; on the recommendations for research pp. 209 ff.

73. A.-K. Wöbse, *Weltnaturschutz: Umweltdiplomatie in Völkerbund und Vereinten Nationen, 1920–1950* (Frankfurt: Campus, 2011). The global environmental movement and its Western origins were mirrored in the setup of international organizations like the WWF and UNEP, and in the International Union for Conservation of Nature (IUCN) that was founded in 1948 as the world's first global environmental organization. The establishment of the MAB programme needs to be seen in relation to decolonization after World War II. Colonial conservation projects continued in the workings of the new international institutions and programmes establishing nature reserves in former colonies.

74. MAB-UNESCO Task Force, *Criteria and Guidelines for the Choice and Establishment of Biosphere Reserves*, Program on Man and the Biosphere (MAB), Final Report (Paris: UNESCO, 1974), p. 2. The guidelines outlined three main categories: natural areas representative of the world's biomes and their subdivisions; unique areas or areas with particular natural features of exceptional interest; and man-modified landscapes.

75. By 1977 118 reserves existed in twenty-seven countries. See W. A. Adams, *Green Development: Environment and Sustainability in the Third World* (London and New York: Routledge, 1992), pp. 33–6.

76. MAB-UNESCO Task Force, *Criteria and Guidelines*, p. 2.

77. The Stockholm conference was the first in the series of UN environmental conferences, continued by the Rio Summit or Earth Summit, the United Nations Conference on Environment and Development in Rio de Janeiro, Brazil, in 1992, and followed by the Earth Summit in 2002, the World Summit on Sustainable Development in Johannesburg, South Africa. J. McCormick *The Global Environmental Movement: Reclaiming Paradise* (London: Belhaven Press, 1989), ch. 5: 'The Stockholm Conference (1970–72)', pp. 88–105, describes how the representatives of the developing countries turned the planned environmental summit into a development summit.

78. Ward and Dubos, *Only One Earth*, p. xviii.

79. Ward and Dubos, *Only One Earth*, p. xviii.

80. Ward and Dubos, *Only One Earth*, p. xviii. Spaceship Earth is spelled in lower case letters and used with the definite article.

81. Ward and Dubos, *Only One Earth*, p. xiii. Jamison holds that '*Only One Earth* made the case for a new kind of environmentalism, combining efficient management of resources with empathetic understanding'. Jamison, *The Making of Green Knowledge*, p. 86.

82. Ward and Dubos, *Only One Earth*, p. xiii (original emphasis).

83. E. B. Weiss, *In Fairness to Future Generations: International Law, Common Patrimony, and Intergenerational Equity* (Tokyo and Dobbs Ferry, NY: Transnational Publishers, 1989).

84. Marshall McLuhan in 1964, as quoted in *Encyclopedia of World Environmental History*, ed. S. Krech III, J. R. McNeill and C. Merchant (New York and London: Routledge,

2004), vol. 1, p. 356.

85. E. Goldsmith, R. Allen, M. Allaby, J. Davoll and S. Lawrence, *A Blueprint for Survival* (London: Tom Stacey, 1972), foreword by Tom Stacey, pp. 7–8. The American edition was published in the same year, introduced by Paul Ehrlich.

86. On earth system modelling see P. N. Edwards, *A Vast Machine: Computer Models, Climate Data, and the Politics of Global Warming* (Boston, MA: MIT Press, 2010); A. Akera, *Calculating a Natural World: Scientists, Engineers, and Computers during the Rise of U.S. Cold War Research* (Cambridge, MA, and London: MIT Press, 2006); B. P. Bloomfield, *Modelling the World: The Social Constructions of Systems Analysts* (Oxford: Basil Blackwell, 1986); F. Elichirigoity, *Planet Management: Limits to Growth, Computer Simulation, and the Emergence of Global Spaces* (Evanston, IL: Northwestern University Press, 1999).

87. H. T. Odum, *Environment, Power, and Society* (New York: Wiley-Interscience, 1971), p. 11.

88. Jamison, *The Making of Green Knowledge*, ch. 3: 'The Dialectics of Environmentalism', pp. 71 ff. See also Worster, *Nature's Economy*.

89. On the work of the Odum brothers establishing the ecosystem concept as a scientific paradigm in ecology see F. B. Golley, *A History of the Ecosytem Concept in Ecology: More than the Sum of the Parts* (New Haven, CT, and London: Yale University Press, 1993), pp. 61–108. Following Golley (p. 105), the new paradigm was formulated most clearly by Eugene Odum in his book on *Fundamentals of Ecology* (Philadelphia, PA: W. B. Saunders, 1953). Compare S. E. Kingsland, *The Evolution of American Ecology, 1890–2000* (Baltimore, MD: Johns Hopkins University Press, 2005); P. Taylor, 'Technocratic Optimism, H. T. Odum, and the Partial Transformation of Ecological Metaphor after World War II', *Journal of the History of Biology*, 21 (1988), pp. 213–44.

90. Accordingly, Jacques Grinevald noted in 1998: 'A scientific consensus on the term [biosphere] is still lacking'. See his introduction to Vernadsky, *The Biosphere*, p. 22. The term and concept of 'ecosystem' was introduced in 1935 by the British botanist Arthur George Tansley; see A. G. Tansley, 'The Use and Abuse of Vegetational Concepts and Terms', *Ecology*, 16 (1935), pp. 284–307; Golley, *A History of the Ecosytem Concept in Ecology*, ch. 2: 'The Genesis of a Concept', pp. 8–34. Compare N. Polunin, 'Our Use of "Biosphere", "Ecosystem" and now "Ecobiome"', *Environmental Conservation*, 11 (1984), p. 198. Nicholas Polunin, a British plant geographer and ecologist, attempted a clarification and unification of the terms in 1984. He defined the 'ecosystem' as '*The total components of an immediate environment or recognizable habitat, including both the inorganic and dead parts of the system and the various organisms which live together in it as a social unit, so far as its characteristic dominance or influence-sphere extends*' (original emphasis). Polunin stressed that his definition deviated from Tansley's insofar as 'spatial delimitation' became 'a necessary refinement' to its meaning.

91. Meadows et al., *The Limits to Growth*. The study and the blueprints it designed met a wide academic and popular reception, and also much criticism, see for instance H. S. D. Cole, C. Freeman, M. Jahoda and K. L. R. Pavitt (eds), *Models of Doom: A Critique of* The Limits to Growth, with a Reply by the Authors of *The Limits to Growth* (New York: Universe Books, 1973). The questions the study raised for consideration were also discussed in *Playboy* magazine: 'A New Set of National Priorities: Three Blueprints for Postwar Reconciliation and Reconstruction', *Playboy,* 18 (1971), pp. 146 ff. For a criticism of the visions of systems operations, including ecological technocracy, see, for instance, H. Marcuse, *One-Dimensional Man: Studies in the Ideology of Advanced*

Industrial Society (Boston, MA: Beacon Press, 1964). On the impact of the study see P. Kupper, '"Weltuntergangs-Visionen aus dem Computer." Zur Geschichte der Studie "Die Grenzen des Wachstums" von 1972', in F. Uekötter and J. Hohensee (eds), *Wird Kassandra heiser? Die Geschichte falscher Ökoalarme* (Stuttgart: Franz Steiner Verlag, 2004), pp. 98–111; T. W. Luke, 'Worldwatching at the Limits of Growth', in T. W. Luke, *Ecocritique: Contesting the Politics of Nature, Economy, and Culture* (Minneapolis, MN, and London: University of Minnesota Press, 1997), pp. 75–94; M. Schoijet, '*Limits to Growth* and the Rise of Catastrophism', *Environmental History*, 4 (1999), pp. 515–30.

92. Meadows et al., *The Limits to Growth*, p. 104.
93. J. W. Forrester, *World Dynamics* (Cambridge, MA: Wright-Allen Press, 1971).
94. Meadows et al., *The Limits to Growth*, p. 123. Three secondary variables, 'crude birth rate', 'crude death rate' and 'services per capita', were integrated into the model.
95. Meadows et al., *The Limits to Growth*, pp. 124, 125.
96. Meadows et al., *The Limits to Growth*, p. 123.
97. Meadows et al., *The Limits to Growth*, p. 127.
98. G. Bowker, 'How to be Universal: Some Cybernetic Strategies, 1943–1970', *Social Studies of Science*, 23 (1993), pp. 107–27, on p. 108; P. N. Edwards, 'The World in a Machine: Origins and Impacts of Early Computerized Global Systems Models', in A. C. Hughes and T. P. Hughes (eds), *Systems, Experts, and Computers: The Systems Approach in Management and Engineering, World War II and After* (Cambridge, MA: MIT Press, 2000), pp. 221–53, on p. 242.
99. Meadows et al., *The Limits to Growth*, ch. V: 'The State of Global Equilibrium', pp. 156 ff.
100. M. Gabel, *Energy, Earth and Everyone: Energy Strategies for Spaceship Earth*, with the World Game Laboratory, Foreword by R. Buckminster Fuller, Afterword by Stewart Brand (1974; New York: Doubleday, 1980), p. 246.
101. M. Gabel, 'Buckminster Fuller and the Game of the World', incl. 'The World Game – How to Make the World Work' from *Utopia or Oblivion*, in *Buckminster Fuller: Anthology for the New Millennium*, ed. T. T. K. Zung (New York: St Martin's Press, 2001), pp. 122–7, 128–31.
102. P. Galison, 'The Ontology of the Enemy: Norbert Wiener and the Cybernetic Vision', *Critical Inquiry*, 21 (1994), pp. 228–66. On the history of cybernetics see also M. Hagner and E. Hörl (eds), *Die Transformation des Humanen: Beiträge zur Kulturgeschichte der Kybernetik* (Frankfurt: Suhrkamp, 2008); S. J. Heims, *The Cybernetics Group* (Cambridge, MA, and London: MIT Press, 1991); Schmidt-Gernig, 'The Cybernetic Society'; C. Pias (ed.), *Cybernetics/Kybernetik: The Macy-Conferences 1946–1953* (Zürich and Berlin: Diaphanes, Vol. 1: *Transactions/Protokolle* (2003); Vol. 2: *Essays & Documents/Essays und Dokumente* (2004)); A. Pickering, *The Cybernetic Brain: Sketches of Another Future* (Chicago, IL: University of Chicago Press, 2009).
103. The original world model underlying *The Limits to Growth* was renewed with the thirty-year update of the Club of Rome's report. *World3-03* delivered similar scenarios to the ones of its predecessor; see D. H. Meadows, J. Randers and D. Meadows, *Limits to Growth: The 30-Year Update* (White River Junction, VT: Chelsea Green Publishing Company, 2004).
104. Meadows et al., *The Limits to Growth*, p. 105; on uncertainty p. 121.
105. P. Erickson, J. L. Klein, L. Daston, R. Lemov, T. Sturm and M. D. Gordin, *How Reason Almost Lost Its Mind: The Strange Career of Cold War Rationality* (Chicago, IL:

University of Chicago Press, 2013); W. P. McCray, *The Visioneers: How a Group of Elite Scientists Pursued Space Colonies, Nanotechnologies, and a Limitless Future* (Princeton, NJ, and Oxford: Princeton University Press, 2013).

106. P. N. Edwards, 'Global Climate Science, Uncertainty and Politics: Data-laden Models, Model-filtered Data', *Science as Culture*, 8 (1999), pp. 437–72.

107. P. J. Taylor and F. H. Buttel, 'How Do We Know We Have Global Environmental Problems? Science and the Globalization of Environmental Discourse', *Geoforum*, 23 (1992), pp. 405–16.

108. W. Sachs, 'Astronautenblick – Über die Versuchung zur Weltsteuerung in der Ökologie', in *Jahrbuch Ökologie 1999* (München: C. H. Beck, 1998), pp. 199–206, on p. 204. James Watt's steam engine was equipped with a 'governor', a contraption that since the late eighteenth century automatically oversaw and controlled the work of a machine, which is to say, observed and regulated a larger production cycle through automated feedback. Sachs points out how the metaphor of Plato's 'ship of state' and its helmsman kept providing a rational basis to call for the 'command of the competent', tying political authority to expertise in an elitist fashion. Environmental governance that is linked both to elite expertise and to an unbending set of rules inscribed into a system's servomechanisms may well take the form of 'ecocracy'. Ibid., p. 205 (translations are mine).

109. Buckminster Fuller, *Operating Manual*, ch. 5: 'General Systems Theory', pp. 57–74, on p. 60.

110. Buckminster Fuller, *Operating Manual*, p. 59.

111. Buckminster Fuller, *Operating Manual*, ch. 3: 'Comprehensively Commanded Automation', pp. 33–47, on p. 46.

112. Buckminster Fuller, *Operating Manual*, ch. 4: 'Spaceship Earth', pp. 48–56, on p. 50.

113. Buckminster Fuller, *Operating Manual*, p. 52.

114. Buckminster Fuller, *Operating Manual*, pp. 52, 54.

115. 'Biosphere (with capitalization also of the initial letter of "The" (*sic*) when immediately preceding it): *The integrated living and life-supporting system comprising the peripheral envelope of Planet Earth together with its surrounding atmosphere so far down, and up, as any form of life exists naturally*'. Polunin, 'Our Use of "Biosphere"' (original emphasis).

116. Vernadsky, *The Biosphere*, § 14, p. 52.

117. J. B. Hagen, *An Entangled Bank: The Origins of Ecosystem Ecology* (New Brunswick, NJ: Rutgers University Press, 1992), Epilogue, 'The Flights of Apollo', pp. 189–98.

118. Hagen, *An Entangled Bank*, p. 190.

119. Hagen, *An Entangled Bank*, p. 190.

120. G. M. Woodwell, 'The Energy Cycle of the Biosphere', in *The Biosphere. A Scientific American* Book (San Francisco, CA: W. H. Freeman and Company, 1970), pp. 26–36, on p. 26.

121. Ehrlich and Harriman, *How to Be a Survivor*.

122. Ehrlich, *The Population Bomb*, p. 132.

4 Storage: The Lifeboats of Human Ecology

1. *Soylent Green* (USA: Warner Brothers/Metro-Goldwyn-Mayer, 1973), directed by Richard Fleischer, starring Charlton Heston, Leigh Taylor Young and Edward G. Robinson. The 2003 DVD's supplementary material includes an extra audio track from 2003 with the retrospective commentaries by director Richard Fleischer and by Leigh

Taylor Young who played the figure of Shirl. The quotes are taken from Fleischer's comments.

2. H. Harrison, *Make Room! Make Room!* (1966; New York: Berkley Publishing Corporation, 1973).

3. F. Osborn, *Our Plundered Planet* (1948; Boston, MA: Little, Brown & Co., 1950); Osborn, *The Limits of the Earth*; F. Osborn, *Our Crowded Planet: Essays on the Pressures of Population* (Garden City, NY: Doubleday, 1962); K. Sax, *Standing Room Only: The World's Exploding Population* (Boston, MA: Beacon Press, 1960) [orig. 1955 *Standing Room Only: The Challenge of Overpopulation*]; Ehrlich, *The Population Bomb*; G. Borgstrom, *Too Many: A Study of Earth's Biological Limitations* (London: Macmillan, 1969).

4. T. Malthus, *An Essay on the Principle of Population, as It Affects the Future Improvement of Society with Remarks on the Speculations of Mr. Godwin, M. Condorcet, and Other Writers* (London, 1798).

5. J. H. Fremlin, 'How Many People Can the World Support?', *New Scientist*, 415 (29 October 1964), pp. 285–7.

6. Ehrlich, *The Population Bomb*, p. 16.

7. Ehrlich, *The Population Bomb*, Foreword by David Brower, p. 14.

8. Ehrlich, *The Population Bomb*, Prologue, p. 11.

9. In recent years, historical studies on demography and population policies after 1945 have proliferated and shifted the focus from population as a statistical construct of the nineteenth century or as an object of racist anthropological sciences in Europe and particularly Germany throughout the Weimar years and the period of National Socialism. These studies also expanded the historiographical perception of population dynamics by focusing neither exclusively on antinatalist eugenics nor on the pronatalist population politics of ambitious nations. Rather, they show that shrinking national populations as well as population growth have sparked discourses of a dying nation for the last 200 years up to the present. See, for instance, A. Bashford, *Global Population: History, Geopolitics and Life on Earth* (New York: Columbia University Press, 2014); M. Connelly, *Fatal Misconception: The Struggle to Control World Population* (Cambridge, MA, and London: Belknap Press of Harvard University Press, 2008).

10. S. Höhler, 'The Law of Growth: How Ecology Accounted for World Population in the 20th Century', *Distinktion. Scandinavian Journal of Social Theory*, 14 (2007), special issue 'Bioeconomy', pp. 45–64.

11. B. Duden, 'Population', in W. Sachs (ed.), *The Development Dictionary: A Guide to Knowledge as Power* (London: Zed Books, 1992), pp. 146–57, on pp. 153–4.

12. On neo-Malthusianism in post-war ecological thought and practice, with a focus on scientific reasoning, see B.-O. Linnér, *The Return of Malthus: Environmentalism and Post-war Population-Resource Crises* (Leverburgh: White Horse Press, 2003); P. Neurath, *From Malthus to the Club of Rome and Back: Problems of Limits to Growth, Population Control, and Migrations* (Armonk, NY: M. E. Sharpe, 1994); Robertson, *The Malthusian Moment*.

13. L. R. Brown, *The Twenty-Ninth Day: Accommodating Human Needs and Numbers to the Earth's Resources* (A Worldwatch Institute Book) (New York: W. W. Norton, 1978), p. 12. According to Lester Brown it was only since the 1977 UN conference on desertification that 'carrying capacity' moved from biology into broader, global vocabulary.

14. Foucault, 'Of Other Spaces', p. 23.

15. Foucault, 'Of Other Spaces', pp. 22, 23.

16. Foucault, *Security, Territory, Population*; Foucault, *The Birth of Biopolitics*; S. Legg,

'Foucault's Population Geographies: Classifications, Biopolitics and Governmental Spaces', *Population, Space and Place*, 11 (2005), pp. 137–56.

17. On the relationship between Foucauldian biopolitics and bioeconomy see L. T. Larsen, 'Speaking Truth to Biopower: On the Genealogy of Bioeconomy', *Distinktion. Scandinavian Journal of Social Theory*, special issue 'Bioeconomy', 14 (2007), pp. 9–24.

18. Duden, 'Population', pp. 146, 149.

19. On 'social physics' and the rise of population statistics in the nineteenth century see A. Desrosières, *The Politics of Large Numbers: A History of Statistical Reasoning* (Cambridge, MA: Harvard University Press, 1998); T. M. Porter, *Trust in Numbers: The Pursuit of Objectivity in Science and Public Life* (Princeton, NJ: Princeton University Press, 1995); T. M. Porter, *The Rise of Statistical Thinking, 1820–1900* (Princeton, NJ: Princeton University Press, 1986).

20. Duden, 'Population', p. 146.

21. S. E. Kingsland, *Modeling Nature: Episodes in the History of Population Ecology* (1985; Chicago, IL, and London: University of Chicago Press, 1995), p. 50.

22. Duden, 'Population', p. 148.

23. On quantification ideals and accounting practices particularly in the nineteenth century see J. L. Klein and M. S. Morgan (eds), *The Age of Economic Measurement* (Durham, NC: Duke University Press, 2001); T. M. Porter, 'The Culture of Quantification and the History of Public Reason', *Journal of the History of Economic Thought*, 26 (2004), pp. 165–77; T. M. Porter, 'The Mathematics of Society: Variation and Error in Quetelet's Statistics', *British Journal for the History of Science*, 18 (1985), pp. 51–69.

24. Sax, *Standing Room Only*, p. xii. This idea goes back to E. A. Ross, *Standing Room Only?* (New York: Century, 1927).

25. Meadows et al., *The Limits to Growth*, p. 29; Brown, *The Twenty-Ninth Day*, introduction p. 1 and back cover.

26. On 'normalization' as a process of producing and applying statistics-based and induced knowledge to which subjects compare and conform to align with their statistical collective, and on the differences between 'normality', 'norm' and 'normativity' see J. Link, *Versuch über den Normalismus: Wie Normalität produziert wird*, 2nd rev. edn (Opladen and Wiesbaden: Westdeutscher Verlag, 1999); W. Sohn and H. Mehrtens (eds), *Normalität und Abweichung: Studien zur Geschichte und Theorie der Normalisierungsgesellschaft* (Opladen, Wiesbaden: Westdeutscher Verlag, 1999); I. Hacking, *The Taming of Chance* (1990; Cambridge: Cambridge University Press, 2004), ch. 19: 'The normal state', pp. 160–9.

27. On visual representation in science and the power of visual images for the perception of reality see, for instance, D. Gugerli and B. Orland (eds), *Ganz normale Bilder: Historische Beiträge zur visuellen Herstellung von Selbstverständlichkeit* (Zürich: Chronos, 2002); B. Heintz and J. Huber (eds), *Mit dem Auge denken: Strategien der Sichtbarmachung in wissenschaftlichen und virtuellen Welten* (Wien and New York: Springer, 2001); M. Lynch and S. Woolgar (eds), *Representation in Scientific Practice* (Cambridge, MA, and London: MIT Press, 1990).

28. Forrester, *World Dynamics*, p. 3.

29. Forrester, *World Dynamics*, p. 3.

30. Kingsland, *Modeling Nature*, ch. 3: 'The Quantity of Life', pp. 50 ff., on p. 56.

31. Kingsland, *Modeling Nature*, p. 56. On the work of Raymond Pearl see especially pp. 56–63 and 'The Logistic Hypothesis', pp. 64 ff. On Alexander Carr-Saunders, Charles Elton and the field of population ecology of the 1920s see P. Anker, *Imperial Ecology:*

Environmental Order in the British Empire, 1895–1945 (Cambridge, MA, and London: Harvard University Press, 2001).

32. R. Pearl, *The Biology of Population Growth* (1925; New York: Alfred A. Knopf, 1930), pp. 28–9.

33. 'Gesetz des Minimums'; J. von Liebig, *Die Chemie in ihrer Anwendung auf Agricultur und Physiologie*, pt 2: *Die Naturgesetze des Feldbaues* (Braunschweig: Vieweg, 1865), pp. 223 ff. (translations are mine).

34. Liebig, *Die Chemie in ihrer Anwendung auf Agricultur und Physiologie*, pt 2, p. 223. Liebig's can be read as an early proposition of a sustainable agricultural economy. See E. Pawley, '*The Balance-Sheet of Nature': Calculating the New York Farm, 1820–1860*, PhD Dissertation Dickinson College (ProQuest, UMI Dissertation Publishing, 2011).

35. Kingsland, *Modeling Nature*, pp. 50–76.

36. Kingsland, *Modeling Nature*, p. 64, on p. 61.

37. T. M. Porter, *Karl Pearson: The Scientific Life in a Statistical Age* (Princeton, NJ: Princeton University Press, 2004). On eugenic traditions in Pearl's work see G. E. Allen, 'Old Wine in New Bottles: From Eugenics to Population Control in the Work of Raymond Pearl', in K. R. Benson, J. Maienschein and R. Rainger (eds), *The Expansion of American Biology* (New Brunswick, NJ, and London: Rutgers University Press, 1991), pp. 231–61; E. Ramsden, 'Carving up Population Science: Eugenics, Demography and the Controversy over the "Biological Law" of Population Growth', *Social Studies of Science*, 32 (2002), pp. 857–99.

38. Kingsland, *Modeling Nature*, p. 58. On Pearl's notion of a law see p. 68. According to Pearl, the Malthusian or exponential population growth curve could not be called a law in the strong sense, since it would eventually lead to absurd values.

39. Pearl, *The Biology of Population Growth*, p. 4. Pearl refers to Verhulst's curve, which he called the 'logistic curve' as early as 1838. In his *Treatise on Man* of 1835, Verhulst's mentor Adolphe Quételet, with reference to Malthus's observations, had proposed two fundamental principles in the analysis of development of population: '*Population tends to increase in a geometrical ratio. The resistance, or sum of the obstacles to its development, is, all things being equal, as the square of the rapidity with which it tends to increase*'. A. Quételet, *A Treatise on Man and the Development of his Faculties*. Facsimile Reproduction of the English Translation of 1842 (New York: Burt Franklin, 1968 [orig. *Sur L'Homme, et le Développement des ses Facultés*, Paris 1835]), p. 49, his emphasis. According to Kingsland, Verhulst was unable to determine the exact nature of the function that described the obstacles to growth. He suggested the obstacles were in *linear* proportion to the size of the 'superabundant' population (the number in excess of the 'normal' population, the one in accord with the resources available). Following Kingsland, in nineteenth-century French the term 'logistique' referred to the art of calculation as opposed to theoretical arithmetic. She suggests that the term connoted 'the idea of a calculating device, from which one could calculate the saturation level of a population and the time when it would reach that level'. Kingsland, *Modeling Nature*, p. 65.

40. Kingsland, *Modeling Nature*, p. 64.

41. Pearl, *The Biology of Population Growth*, pp. 3, 18.

42. Kingsland, *Modeling Nature*, pp. 67–9.

43. Kingsland, *Modeling Nature*, p. 75.

44. M. Sanger (ed.), *Proceedings of the World Population Conference, held at the Salle Centrale, Geneva, August 29ᵗʰ to September 3ʳᵈ, 1927* (London: Edward Arnold & Co.,

1927), p. 5.

45. M. Connelly, 'To inherit the Earth. Imagining world population, from the yellow peril to the population bomb', *Journal of Global History*, 1 (2006), pp. 299–319, on p. 304.

46. Duden, 'Population', p. 149.

47. Duden, 'Population', p. 150.

48. W. M. Chamberlain, 'Population Control: The Legal Approach to a Biological Imperative', *California Law Review*, 58 (1970), pp. 1414–43, on p. 1414. See also K. Davis, 'Population Policy: Will Current Programs Succeed?', *Science*, 158 (1967), pp. 730–9; P. R. Ehrlich and A. H. Ehrlich, 'Population Control and Genocide', in J. P. Holdren and P. R. Ehrlich (eds), *Global Ecology: Readings Toward a Rational Strategy for Man* (New York: Harcourt Brace Jovanovich, 1971), pp. 157–9; P. R. Ehrlich and J. P. Holdren, 'Impact of Population Growth', *Science* (1971), pp. 1212–17; S. Enke, 'Birth Control for Economic Development', in Holdren and Ehrlich (eds), *Global Ecology*, pp. 193–200.

49. P. T. Piotrow, *World Population Crisis: The United States Response* (New York: Praeger, 1973), p. vii.

50. Chamberlain, 'Population Control', p. 1414 (original emphasis).

51. Chamberlain, 'Population Control', footnote 2, p. 1414.

52. R. M. Salas, *International Population Assistance: The First Decade. A Look at the Concepts and Politics which have Guided the UNFPA in its First Ten Years* (New York: Pergamon Press, 1979), p. 125.

53. Duden, 'Population', p. 149.

54. Ehrlich et al., *Ecoscience*; P. R. Ehrlich and A. H. Ehrlich, *Population, Resources, Environment: Issues in Human Ecology* (San Francisco, CA: W. H. Freeman, 1970 and 1972).

55. Osborn, *The Limits of the Earth*, p. 77.

56. Osborn, *The Limits of the Earth*, p. 207 (original emphasis).

57. Osborn, *Our Plundered Planet*, p. 43.

58. U. Pörksen, 'Logos, Kurven, Visiotype', in U. Gerhard, J. Link and E. Schulte-Holtey (eds), *Infografiken, Medien, Normalisierung: Zur Kartographie politisch-sozialer Landschaften* (Heidelberg: Synchron, 2001), pp. 63–76. For a definition of 'visiotype' see p. 67, for its characteristics p. 73, for visual 'trends' p. 67.

59. Pörksen, 'Logos, Kurven, Visiotype', p. 74.

60. Pörksen, 'Logos, Kurven, Visiotype', p. 74. Pörksen refers only to the (empirical) exponential population growth curve, not to the modelled S-curve with its upper part of regression.

61. Pörksen, 'Logos, Kurven, Visiotype', p. 74. On the power of curves to pull people into communities, for example communities of fear, see p. 67.

62. Pörksen, 'Logos, Kurven, Visiotype', pp. 73–4.

63. Ehrlich, *The Population Bomb*, front cover.

64. Meadows et al., *The Limits to Growth*, pp. 91–2.

65. Forrester, *World Dynamics*; Edwards, 'The World in a Machine', pp. 236–40.

66. Edwards, 'The World in a Machine', p. 238.

67. Meadows et al., *The Limits to Growth*, p. 125.

68. Link, *Versuch über den Normalismus*, p. 200; see also J. Link, 'Aspekte der Normalisierung von Subjekten', in U. Gerhard, J. Link and E. Schulte-Holtey (eds), *Infografiken, Medien, Normalisierung: Zur Kartographie politisch-sozialer Landschaften* (Heidelberg: Synchron, 2001), pp. 77–92, on p. 88, for a typology of basic normalistic

curves (growth curves, regression curves, and S-curves) and corresponding normalistic narrations.

69. Meadows et al., *The Limits to Growth*, p. 156.
70. Sax, *Standing Room Only*, p. 177.
71. Sax, *Standing Room Only*, p. 11. The corresponding study Sax points to is A. Myrdal and P. Vincent, *Are there Too Many People?* (New York: Manhattan Publishing Company, 1950). Malthus had outlined the two options of 'positive checks' and 'negative checks' to population growth in his work *An Essay on the Principle of Population* (1798).
72. See K. Davis, 'Zero Population Growth: The Goal and the Means', in M. Olson and H. H. Landsberg (eds), *The No-Growth Society* (1973; London: Woburn, 1975), pp. 15–30. Davis is said to have coined the term 'zero population growth' in 1967.
73. K. E. Boulding, *The Meaning of the Twentieth Century: The Great Transition* (1964; New York: Harper & Row, 1965), ch. 6: 'The Population Trap', pp. 121–36; on the 'child certificates' pp. 135–6.
74. Ehrlich, *The Population Bomb*, front cover.
75. Ehrlich, *The Population Bomb*, see p. 4 for a reference to Moore's pamphlet.
76. Ehrlich, *The Population Bomb*, p. 67.
77. Ehrlich, *The Population Bomb*, p. 149.
78. Ehrlich, *The Population Bomb*, p. 150.
79. Ehrlich, *The Population Bomb*, p. 156. Ehrlich's work is far-sighted where it assesses environmental problems like resource exploitation, fossil energy use, nuclear waste deposits, overfishing and species extinction (pp. 46 ff.). Ehrlich also addresses the problem of climate change due to the 'greenhouse effect', which at his time entered not environmental but scientific discourse. Several of Ehrlich's requests took hold in environmental politics, as for example the concepts of the 'global commons', of 'intergenerational justice', of 'sustainability' and of 'sustainable development'. Some of the instruments he proposed were eventually developed into standards of environmental law, like the 'polluters pay' principle (p. 152), which denotes the monetary compensation of environmental damages according to the principle that the party responsible is liable for the damages (p. 100). Overall, Ehrlich's book is an empathic plea against 'the greed and stubbornness of industries, the recalcitrance of city governments, the weakness of state control, and the general apathy of the American people' (p. 123).
80. Ehrlich, *The Population Bomb*, p. 166.
81. Ehrlich, *The Population Bomb*, pp. 29, 21.
82. Ehrlich, *The Population Bomb*, p. 29: 'some are always outreproducing others'. Malthus had spoken of the 'law of necessity' (of acquiring food) and of the merely arithmetic increase of the means of subsistence that would work as control mechanisms on the geometrical growth of population.
83. Ehrlich, *The Population Bomb*, p. 20. Ehrlich speaks in his book of 'human surplus', 'overpopulation' and 'overcrowding' (p. 167). On the economy of surplus populations see also M. Cooper, *Life as Surplus: Biotechnology and Capitalism in the Neoliberal Era* (Seattle, WA: University of Washington Press, 2008). The concept of an excess population had already been a central element of Malthus's theory in 1798.
84. Ehrlich, *The Population Bomb*, p. 34.
85. Ehrlich, *The Population Bomb*, Prologue, pp. 11, 38. In 1966 US President Lyndon B. Johnson had announced that the United States would lead the world in a war against hunger.

86. Ehrlich, *The Population Bomb*, pp. 69–70, pp. 72 ff.
87. Ehrlich, *The Population Bomb*, p. 79.
88. Ehrlich, *The Population Bomb*, pp. 133–6. It is characteristic of the population discourse of the time to perceive births exclusively under the premise of societal costs and individual benefits.
89. For an analysis of this odd alliance see Robertson, *The Malthusian Moment*.
90. Ehrlich, *The Population Bomb*, p. 166.
91. '30 Years of ZPG', *Reporter*, December 1998, pp. 12–19, *The Population Connection*, at http://www.populationconnection.org/site/DocServer/1219thirtyyears. pdf?docID=261 [accessed 28 August 2014], pp. 13, 17. According to this article, 'the years between 1969 and 1972 saw the membership of ZPG briefly blossom to more than 35,000 members', p. 13.
92. '30 Years of ZPG', pp. 18, 15. In 1969 the US folk musician Pete Seeger wrote a song for ZPG that turned Malthusian geometrical growth into music: The first verse of 'We'll All Be A-Doubling [in forty-two years]' starts out with '2x2 is 4! 2x4 is 8! 2x8 is 16, and the hour is getting late!' Ibid., p. 15.
93. *Z.P.G. Zero Population Growth* (USA: Paramount, 1971), directed by Michael Campus, starring Oliver Reed and Geraldine Chaplin.
94. Ehrlich, *The Population Bomb*, pp. 123, 11, 112.
95. Ehrlich, *The Population Bomb*, p. 132. In the German translation the phrase 'good ship Earth' actually became 'Spaceship Earth' ('Raumschiff Erde'). P. R. Ehrlich, *Die Bevölkerungsbombe* (1971; Frankfurt: Fischer, 1973), p. 86.
96. Ehrlich, *The Population Bomb*, p. 165.
97. Ehrlich, *The Population Bomb*, p. 165.
98. S. Weissman, 'Die Bevölkerungsbombe ist ein Rockefeller-Baby', in H. M. Enzensberger and K. M. Michel (eds), *Ökologie und Politik oder Die Zukunft der Industrialisierung* (*Kursbuch* 33) (Berlin: Kursbuch/Rotbuch Verlag, 1973), pp. 81–94.
99. Ehrlich eventually lost a legendary bet to the economist Julian Simon about the future prices of basic resources. In 1980 Ehrlich and Simon bet on the development of prices of commodity metals within the next decade. While Ehrlich assumed that prices would go up, Simon's prognosis that prices would actually go down turned out to be right. In most cases new technologies involving different materials counterbalanced expected scarcities and increasing costs. Simon held that if resources per capita decreased human performance would improve. See J. L. Simon, *The Ultimate Resource* (Princeton, NJ: Princeton University Press, 1981).
100. See the definition given in G. Hardin, 'Ethical Implications of Carrying Capacity', in G. Hardin (ed.), *Managing the Commons* (San Francisco, CA: Freeman, 1977), pp. 112–25, particularly p. 113.
101. This definition is according to Brown, *The Twenty-Ninth Day*, p. 13.
102. Holdren and Ehrlich, 'Human Population', on p. 288.
103. Ehrlich, *The Population Bomb*, p. 167 (emphasis added). The question of an 'optimum population' had been posed at the time of World War I and the subsequent World Population Conference in 1927. H. P. Fairchild, 'Optimum Population', in M. Sanger (ed.), *Proceedings of the World Population Conference, held at the Salle Centrale, Geneva, August 29th to September 3rd, 1927* (London: Edward Arnold & Co., 1927), pp. 72–85. Based on neo-Malthusian demography, Darwinian evolutionary theory, and Galton's eugenics population ecologist Alexander Carr-Saunders had argued in his 1922 book *The Population Problem* that human populations fluctuated around their

'optimum number'. A. M. Carr-Saunders, *The Population Problem: A Study in Human Evolution* (1922; New York: Arno, 1974).

104. G. Hardin, 'The Tragedy of the Commons', *Science*, 162 (1968), pp. 1243–8, on p. 1248.

105. G. Hardin, 'Carrying Capacity as an Ethical Concept', in G. R. Lucas Jr, and T. W. Ogletree (eds), *Lifeboat Ethics: The Moral Dilemmas of World Hunger* (New York: Harper & Row, 1976), pp. 120–37, on p. 134.

106. Hardin, 'The Tragedy of the Commons', p. 1244.

107. G. Hardin, *Exploring New Ethics for Survival: The Voyage of the Spaceship Beagle* (New York: Viking Press, 1972), pp. 201, 192. See also G. Hardin, 'Editorial: Parenthood: Right or Privilege?', *Science*, 169 (1970), p. 427.

108. S. Höhler, '"The Real Problem of a Spaceship is its People": Spaceship Earth as Ecological Science Fiction', in G. Canavan and K. S. Robinson (eds), *Green Planets: Ecology and Science Fiction* (Middletown, CT: Wesleyan University Press, 2014), pp. 99–114.

109. L. Daston, 'The Moral Economy of Science', in A. Thackray (ed.), *Constructing Knowledge in the History of Science* (*Osiris*, 10 (1995)), pp. 3–24, referring to the British historian E. P. Thompson's use of the term in his work on British social history in the nineteenth and twentieth centuries.

110. Ehrlich, *The Population Bomb*, p. 133.

111. W. Paddock and P. Paddock, *Famine – 1975!* (London: Weidenfeld & Nicolson, 1968) [orig. *Famine – 1975! America's Decision: Who will Survive?* Boston, MA: Little, Brown & Co., 1967], ch. 3, pt 9, pp. 205 ff. With explicit reference to the Paddocks Ehrlich suggested triage as a strategy of 'rational choice'; Ehrlich, *The Population Bomb*, p. 161.

112. Paddock and Paddock, *Famine*, pp. 206–7.

113. Paddock and Paddock, *Famine*, p. 230.

114. G. Hardin, 'Lifeboat Ethics: The Case Against Helping the Poor', *Psychology Today*, 8 (1974), pp. 38–43, 123–6. For an extended version see G. Hardin, 'Living on a Lifeboat', *BioScience*, 24 (1974), pp. 561–8.

115. Hardin, 'Living on a Lifeboat', p. 564.

116. See also P. Næss, 'Live and Let Die: The Tragedy of Hardin's Social Darwinism', *Journal of Environmental Policy & Planning*, 6 (2004), pp. 19–34.

117. Hardin, 'Living on a Lifeboat', p. 565.

118. Hardin, 'Living on a Lifeboat', p. 564 (original emphasis).

119. Link, *Versuch über den Normalismus*, p. 200. Following Hardin, this route of irreversible 'denormalization' had to be controlled for the benefit and stability of the donor countries.

120. Hardin, 'Living on a Lifeboat', p. 564. Notably, the 'normal' in Hardin's model refers to nations unable to economize well, which will consequently go extinct.

121. Nature scenes in *Soylent Green* were taken from *National Geographic* material.

122. Holdren and Ehrlich, 'Human Population', p. 283.

123. In response to rapid population growth after World War II, Stanford University, the Carnegie Institution, the Rockefeller Foundation, and several other US institutions mass-cultivated the single-celled algae *Chlorella* to serve as a source of protein-rich food and energy; see W. Belasco, 'Algae Burgers for a Hungry World? The Rise and Fall of Chlorella Cuisine', *Technology and Culture*, 38 (1997), pp. 608–34.

124. The original formula was presented in 1971 in Ehrlich and Holdren, 'Impact of Population Growth'. A reworked model was presented in 1972. See also G. C. Daily and P.

R. Ehrlich, 'Population, Sustainability, and Earth's Carrying Capacity', *BioScience*, 42 (1992), pp. 761–71, on the IPAT formula see p. 762.

125. E. B. Barbier, J. C. Burgess and C. Folke, *Paradise Lost? The Ecological Economics of Biodiversity* (London: Earthscan, 1994), pp. 44–5.

126. Catton, *Overshoot*, p. 126.

127. A. Agrawal, *Environmentality: Technologies of Government and the Making of Subjects* (New Delhi: Oxford University Press, 2005); T. W. Luke, 'Environmentality as Green Governmentality', in É. Darier (ed.), *Discourses of the Environment* (Oxford: Blackwell, 1999), pp. 121–51.

5 Classification: Biosphere Reserves

1.   *Logan's Run* (USA: Warner Brothers/Metro-Goldwyn-Mayer, 1976), directed by Michael Anderson, starring Michael York, Jenny Agutter and Peter Ustinov. The movie was based on the novel *Logan's Run* by William F. Nolan and George Clayton Johnson (New York: Dial Press, 1967). I quote from the written prologue introducing the movie.

2.   Sloterdijk, *Sphären*, pp. 260–1.

3.   Sloterdijk, *Sphären*, p. 251.

4.   S. Höhler, 'The Environment as a Life Support System: the Case of Biosphere 2', *History and Technology*, 26 (2010), pp. 39–58.

5.   W. Kuhns, *The Post-Industrial Prophets: Interpretations of Technology* (New York: Harper and Row, 1971), ch. 10: 'Leapfrogging the Twentieth Century. R. Buckminster Fuller', pp. 220–46, on p. 222.

6.   K. E. Boulding, 'Spaceship Earth Revisited', in H. E. Daly (ed.), *Economics, Ecology, Ethics: Essays Toward a Steady-State Economy* (San Francisco, CA: W. H. Freeman and Company, 1980), pp. 264–6.

7.   Sloterdijk, *Sphären*, p. 254.

8.   Philip Hawes, the architect of Biosphere 2, as quoted in J. Allen, *Biosphere 2: The Human Experiment* (New York: Penguin Books, 1991), p. 16. Biosphere 2 was also called a 'glass spaceship'; K. Kelly, 'Biosphere 2 at One', *Whole Earth Review*, 77 (1992), pp. 90–105, on p. 105.

9.   Barthes, 'Myth Today', p. 129.

10. Following Barthes, driven either to unveil or to liquidate the concept it presents, myth will 'naturalize' it.

11. J. Rifkin, *Algeny: A New Word – A New World* (Harmondsworth: Penguin, 1983), pt 7: 'Choices', pp. 245 ff., on p. 252.

12. Allen, *Biosphere 2*; J. Allen, *Me and the Biospheres: A Memoir by the Inventor of Biosphere 2* (Santa Fe, NM: Synergetic Press, 2009); A. Alling and M. Nelson, *Life Under Glass: The Inside Story of Biosphere 2* (Oracle, AZ: Space Biospheres Ventures and The Biosphere Press, 1993); R. Reider, *Dreaming the Biosphere: The Theater of All Possibilities* (Albuquerque, NM: University of New Mexico Press, 2009). See the project website 'Biospherics' at http://www.biospheres.com. During the experiment, the *Whole Earth Review*, a journal published by the editors of the *Whole Earth Catalog*, provided regular updates: K. Kelly, 'Biosphere II: An Autonomous World, Ready to Go', *Whole Earth Review*, 67 (1990), pp. 2–13; Kelly, 'Biosphere 2 at One'.

13. Allen, *Biosphere 2*, p. 18. The Institute of Ecotechnics had been established in 1973 to study the ecological 'building blocks' of the earth's biosphere; ibid., p. 2. On the insti-

tute's history see J. Allen, T. Parrish and M. Nelson, 'The Institute of Ecotechnics: An Institute Devoted to Developing the Discipline of Relating Technosphere to Biosphere', *Environmentalist*, 4 (1984), pp. 205–18.

14. R. Buckminster Fuller, 'Noah's Ark #2' [facsimile reproduction of an unpublished manuscript titled 'Project Noah's Ark #2', written in 1950], in: J. Krausse and C. Lichtenstein (eds), *Your Private Sky: Richard Buckminster Fuller, The Art of Design Science* (Baden: Lars Müller, 1999), pp. 176–226.

15. 'The cost of snow removal in New York City would pay for the dome in 10 years'. R. Buckminster Fuller, 'The Case for a Domed City', *St. Louis Post Dispatch* (26 September 1965), pp. 39–41.

16. On spaceframe design, sealing, pressure equilibration and cooling see Allen, *Biosphere 2*, ch. 5: 'Technics', pp. 59–68. On energy consumption, B. Zabel, P. Hawes, H. Stuart and B. D. V. Marino, 'Construction and Engineering of a Created Environment: Overview of the Biosphere 2 Closed System', in B. D. V. Marino and H. T. Odum (eds), *Biosphere 2: Research Past and Present* (Amsterdam: Elsevier Science, 1999), pp. 43–63, table 'Energy center facts for Biosphere 2', p. 60. Yearly consumption of Biosphere 2 came to an average of 4 to 5 million kilowatt-hours.

17. J. Allen and M. Nelson, *Space Biospheres* (1986; Oracle, AZ: Synergetic Press, 1989), p. 21.

18. The image 'Major Cycles of the Biosphere' from G. Evelyn Hutchinson's *Scientific American* issue on the Biosphere (1970) was reprinted in Allen, *Biosphere 2*, p. 7.

19. L. Gentry and K. Liptak, *The Glass Ark: The Story of Biosphere 2* (New York: Puffin Books (Viking Penguin), 1991), p. 82.

20. Allen, *Biosphere 2*, pp. 95–6 on 'Microcity', 'Micropolis' and 'Habitat'. On 'intensive agriculture' see ch. 6: 'The Farm', pp. 73–87.

21. R. F. Dasmann, 'Toward a Biosphere Consciousness', in D. Worster (ed.), *The Ends of the Earth: Perspectives on Modern Environmental History* (Cambridge: Cambridge University Press, 1988), pp. 277–88.

22. On scaling up laboratory conditions to universalize laboratory-generated knowledge see B. Latour, 'Give Me a Laboratory and I Will Raise the World', in K. Knorr-Cetina and M. Mulkay (eds), *Science Observed: Perspectives in the Social Studies of Science* (London: Sage, 1983), pp. 141–70.

23. Allen, *Biosphere 2*, ch. 2: 'The Quantum Leap', pp. 21–31; on the sealed 'test module' pp. 22–4, on p. 24.

24. Allen, *Biosphere 2*, p. 95.

25. Allen, *Biosphere 2*, p. 95.

26. Allen, *Biosphere 2*, pp. 26–31 on 'Human Experiments'.

27. Mission One was the first experimental closure of Biosphere 2 with eight crew members between 1991 and 1993. J. Allen and M. Nelson, 'Biospherics and Biosphere 2, Mission One (1991–93)', in B. D. V. Marino and H. T. Odum (eds), *Biosphere 2: Research Past and Present (Ecological Engineering*, 13: 1–4 (1999), Special Issue 'Biosphere 2') (Amsterdam: Elsevier Science, 1999), pp. 15–29. Mission 2 commenced in March 1994 and was terminated after six months. See B. D. V. Marino and H. T. Odum, 'Biosphere 2: Introduction and Research Progress', in Marino and Odum (eds), *Biosphere 2*, pp. 3–14.

28. Allen, *Biosphere 2*, p. 2 (emphasis added).

29. Allen and Nelson, *Space Biospheres*, preface, p. ii.

30. Allen, *Biosphere 2*, pp. 1, 59.

31. Allen and Nelson, *Space Biospheres*, p. 55.

32. Allen, *Biosphere 2*, p. 3. On 'cabin ecology' see P. Anker, 'The Ecological Colonization of Space', *Environmental History*, 10 (2005), pp. 239–68; P. Anker, 'The Closed World of Ecological Architecture', *Journal of Architecture*, 10 (2005), pp. 527–52.

33. Allen and Nelson, *Space Biospheres*, ch. 3: 'Mars Settlement', pp. 65–81, on p. 66.

34. On plans to use the capacity of the Saturn-V rocket in developing a space station, initiated and promoted by Wernher von Braun during the 1960s, see T. A. Heppenheimer, *The Space Shuttle Decision: NASA's Search for a Reusable Space Vehicle*, National Aeronautics and Space Administration (Washington, DC: NASA History Office, 1999), pp. 60 ff.

35. R. E. Bilstein, *Orders of Magnitude: A History of NACA and NASA, 1915–90*, National Aeronautics and Space Administration (Washington, DC: Scientific and Technical Information Division, 1989), p. 108 on the report of Space Task Group to the President. See Heppenheimer, *The Space Shuttle Decision*, ch. 3: 'Mars and Other Dream Worlds', pp. 105 ff., for NASA deliberations during the 1960s regarding Mars and on the significance of a reusable shuttle. See also A. J. Butrica, *The Navigators: A History of NASA's Deep-Space Navigation* (CreateSpace Publishing, 2014); T. Hogan, *Mars Wars: The Rise and Fall of the Space Exploration Initiative*, National Aeronautics and Space Administration (Washington, DC: NASA History Division, Office of External Relations, 2007); H. E. McCurdy, *The Space Station Decision: Incremental Politics and Technological Choice* (Baltimore, MD: Johns Hopkins University Press, 1990). For the full text version of the report of the Space Task Group see 'The Post-Apollo Space Program: Directions for the Future, September 1969', National Aeronautics and Space Administration, NASA History Division, at http://www.hq.nasa.gov/office/pao/History/taskgrp.html [accessed 28 August 2014].

36. Heppenheimer, *The Space Shuttle Decision*, ch. 2: 'NASA's Uncertain Future', pp. 55 ff.

37. Space probes Viking 1 and Viking 2 were launched on 20 August and 9 September 1975. The twin mission reached Mars orbit in the summer of 1976. Each probe consisted of two parts. While the orbiters collected and processed data of Mars to Earth, the landers reached the planet and transmitted the first close photographic images from the Martian surface.

38. 'President Nixon's Announcement on the Development of the Space Shuttle, January 5, 1972', Key Documents in the History of Space Policy, National Aeronautics and Space Administration, NASA History Division, at http://www.hq.nasa.gov/office/pao/History/spdocs.html [accessed 28 August 2014].

39. 'President Nixon's Announcement on the Development of the Space Shuttle'.

40. 'President Nixon's Announcement on the Development of the Space Shuttle', supplement statement by NASA administrator James Fletcher on the statement by President Nixon. The reusable Space Shuttle, realized in the 1980s, offered a cheaper routine vehicle for more and shorter missions. Heppenheimer subsumes the building of the shuttle under 'Low-Cost Reusable Space Flight'; Heppenheimer, *The Space Shuttle Decision*, p. 73. See also T. A. Heppenheimer, *Development of the Shuttle, 1972–1981* (Washington, DC, and London: Smithsonian Institution Press, 2002); D. M. Hartland, *The Story of the Space Shuttle* (London et al.: Springer, 2004). First tests with a shuttle spacecraft were made in the 1960s. The *Enterprise* in 1977 was a test vehicle. The first shuttle to fly was *Columbia* on 12 April 1981.

41. 'In retrospect, one can say that this "sell-off" of spaceships and rockets was enormously profitable: With an expenditure of $2.4 billion, more than 29,000 people could be employed for years to come'. B. Stanek, *Raumfahrt Lexikon* (Bern and Stuttgart: Hallwag,

1983), p. 257 (translation is mine). The Saturn-V rocket was reused to house Skylab and to bring the space station into earth orbit. The Orbital Workshop consisted of the third stage of a Saturn-V rocket. Likewise, the mothballed Saturn I-B rockets were reused to bring the crews into orbit. An Apollo spacecraft served as the carrier vehicle. Ibid., pp. 257–63.

42. D. J. Shayler, *Skylab: America's Space Station* (London: Springer, 2001), p. 70.

43. Skylab was launched on 14 May 1973, and stayed in Earth orbit until 1979. The workshop was visited three times in 1973 and 1974. The third and last mission in February 1974 was the longest, lasting eighty-four days. See D. J. Shayler, *Skylab*; W. D. Compton and C. D. Benson, *Living and Working in Space: A History of Skylab* (Washington, DC: National Aeronautics and Space Administration. 1983); P. Baker, *The Story of Manned Space Stations: An Introduction* (London: Springer, 2001). Skylab was the closest the US came to building a habitable station in space until construction of the International Space Station ISS was begun in the late 1990s.

44. The US space programme suffered severely in the months after the explosion of the *Challenger* in 1986, and until 1988 no further shuttle start took place. This gap was also noticeable in the international Spacelab programme of the 1980s and 1990s, as the modular space laboratory housing up to four scientists was carried in the shuttle's cargo bay. Technologically more advanced than Skylab ten years earlier, Spacelab pursued a similar aim of making space accessible through scientific experimentation. Spacelab missions took place between 1983 and 1998. They formed the beginning of a manned European space programme with the long-term goal of participating in a future space station. D. R. Lord, *Spacelab: An International Success Story* (Washington, DC: National Aeronautics and Space Administration, Scientific and Technical Division, 1987).

45. Salyut was the umbrella term for a program of seven Soviet space stations launched between 1971 and 1982. Salyut 1, launched on 19 April 1971, was the prototype of the scientific orbital station. See, for example, S. Haeuplik-Meusburger, C. Paterson, D. Schubert, and P. Zabel, 'Greenhouses and their Humanizing Synergies', *Acta Astronautica*, 96 (2014), pp. 138–50. I do not methodically survey the Soviet space history in this book. A systematic comparison across the Iron Curtain still needs to be achieved. For recent scholarship in this direction see J. T. Andrews and A. A. Siddiqi, *Into the Cosmos: Space Exploration and Soviet Culture* (Pittsburgh, PA: University of Pittsburgh Press, 2011); S. Gerovitch, *From Newspeak to Cyberspeak: A History of Soviet Cybernetics* (Cambridge, MA, and London: MIT Press, 2004); E. Maurer, J. Richers, M. Ruthers and C. Scheide (eds), *Soviet Space Culture: Cosmic Enthusiasm in Socialist Societies* (Houndmills, Basingstoke, and New York: Palgrave Macmillan, 2011).

46. Allen and Nelson, *Space Biospheres*, p. 67.

47. Allen and Nelson, *Space Biospheres*, p. 67.

48. Ronald Reagan in his address on the State of the Union on 25 January 1984, as quoted in R. D. Launius, *Space Stations: Base Camps to the Stars* (Washington, DC, and London: Smithsonian Books, 2003), pp. 120, 121. The space station *Freedom* project was eventually delayed and ultimately abandoned for its too high ambitions. From the plans the International Space Station ISS was developed. The ISS project began in 1998 and the first team moved in during the year 2000.

49. 'Pioneering the Space Frontier'. The Report of the National Commission on Space, May 1986. Key Documents in the History of Space Policy, National Aeronautics and Space Administration, NASA History Division, at http://www.hq.nasa.gov/office/pao/History/spdocs.html [accessed 28 August 2014]. The so-called Paine Report had to face

the Presidential Commission Report subsequent to the *Challenger* accident in 1986, commonly called the Rogers Commission Report, published in 1987.

50. Allen, *Biosphere 2*, p. 33, emphasis in the original.

51. C. Sagan, *Pale Blue Dot: A Vision of the Human Future in Space* (New York: Random House, 1994). The science fiction movie *2001: A Space Odyssey*, directed by Stanley Kubrick (USA: Metro-Goldwyn-Mayer, 1968), featured a torus or wheel-like space station modelled on the von Braun version suggested in the early 1950s, complete with provisions for artificial gravity by rotation and centrifugal force.

52. Whitney Matthews from NASA's Goddard Space Flight Center. W. Matthews, 'Miniaturization for Space Travel', in H. D. Gilbert (ed.), *Miniaturization* (New York: Reinhold Publishing Corporation, 1961), pp. 257–69, on p. 257.

53. Matthews, 'Miniaturization for Space Travel', p. 269.

54. *Die Kolonisierung des Weltraums*, by the editors of *Time-Life Books* (Amsterdam: Time-Life Books, 1991) p. 111 (translation is mine).

55. 'NASA Leadership and America's Future in Space'. A Report to the Administrator by Dr. Sally K. Ride, August 1987. Key Documents in the History of Space Policy, National Aeronautics and Space Administration, NASA History Division, at http://www.hq.nasa.gov/office/pao/History/spdocs.html [accessed 28 August 2014].

56. 'Beyond Earth's Boundaries: Human Exploration of the Solar System in the 21st Century'. 1988 Annual Report to the Administrator, National Aeronautics and Space Administration, Washington, DC, Office of Exploration 1988, p. 48; see *Educational Resources Information Center (ERIC)*, at http://eric.ed.gov:80/ERICDocs/data/eric-docs2sql/content_storage_01/0000019b/80/1f/74/52.pdf [accessed 28 August 2014].

57. 'President Bush's Remarks on the Twentieth Anniversary of the Apollo 11 Moon Landing (his space exploration initiative speech), October 4, 1989', Key Documents in the History of Space Policy, National Aeronautics and Space Administration, NASA History Division, at http://www.hq.nasa.gov/office/pao/History/spdocs.html [accessed 28 August 2014].

58. Allen, *Biosphere 2*, p. 149.

59. Allen, *Biosphere 2*, p. 115.

60. Allen and Nelson, *Space Biospheres*, p. 55.

61. Allen, *Biosphere 2*, p. 99 (original emphasis). The 24th American Astronautical Society (AAS) Goddard Memorial Symposium 'The Human Quest in Space' took place 20–21 March 1986, in Greenbelt, Maryland, USA. The pilot and astronaut Russell (Rusty) Schweickart had been a crewmember of the Apollo 9 mission in 1969.

62. Hawes quoted in Allen, *Biosphere 2*, p. 16 (original emphasis).

63. Allen, *Biosphere 2*, p. 10.

64. E. Schrödinger, *What is Life? The Physical Aspect of the Living Cell* (1944). With *Mind and Matter* and *Autobiographical Sketches* (Cambridge and New York: Cambridge University Press, 1992).

65. J. Monod, *Chance and Necessity: An Essay on the Natural Philosophy of Modern Biology* (New York: Vintage Books, 1971) [orig. French 1970].

66. J. D. Bernal, *The Origin of Life* (London: Weidenfeld & Nicolson, 1967).

67. L. Margulis and D. Sagan, *What is Life?* (New York: Simon & Schuster, 1995).

68. M. Eigen, *Stufen zum Leben: Die frühe Evolution im Visier der Molekularbiologie*, München/Zürich, Piper 1987 [Engl. *Steps towards Life: A Perspective on Evolution*].

69. Allen, *Biosphere 2*, 'The Pioneers', pp. 5–19, on the experiments of the Institute of Biomedical Problems in Moscow and the Institute of Biophysics in Krasnoyarsk, Siberia,

pp. 12–15; Alling and Nelson, *Life Under Glass*, pp. 196–7.

70. R. A. Wharton Jr, D. T. Smernoff and M. M. Averner, 'Algae in Space', in C. A. Lembi and J. R. Waaland (eds), *Algae and Human Affairs* (Cambridge: Cambridge University Press, 1988); F. B. Salisbury, 'Joseph I. Gitelson and the Bios-3 Project', *Life Support & Biosphere Science. International Journal of Earth Space*, 1 (1994), pp. 69–70.

71. Allen, *Biosphere 2*, p. 13. C. E. Folsome and J. A. Hanson, 'The Emergence of Materially Closed System Ecology', in N. Polunin (ed.), *Ecosystem Theory and Application* (New York: John Wiley & Sons Ltd, 1986), pp. 269–88; In the early 1980s bottled eco-spheres were first developed for large-scale marketing to a general public. See also the cover story by P. Warshall, 'The Ecosphere: Introducing an Einsteinian Ecology', *Whole Earth Review*, 46 (1985), pp. 28–31.

72. Allen, *Biosphere 2*, p. 15.

73. Allen, *Biosphere 2*, p. 6: 'In 1969 [1970, S.H.], Scientific American published the influential volume entitled The Biosphere, for which Hutchinson wrote the introduction. This publication introduced the name and work of Vernadsky to John Allen, Mark Nelson and their colleagues who would later conceive of the idea for Biosphere 2'.

74. Allen, *Biosphere 2*, on Vernadsky see pp. 5–6; on the work of James Lovelock and Lynn Margulis and its application to Biosphere 2 see pp. 7–8.

75. Allen, *Biosphere 2*, p. 8.

76. 'G. Evelyn Hutchinson made the same claim about the biosphere almost half a century ago'. Hagen, *An Entangled Bank*, p. 194. In her biography of Odum Betty Jean Craige asserts that 'ecosystem ecology became mainstream ecology' around 1960; B. J. Craige, *Eugene Odum: Ecosystem Ecologist and Environmentalist* (2001; Athens, GA: University of Georgia Press, 2002), p. 79.

77. Allen, *Biosphere 2*, p. 3.

78. Hagen, *An Entangled Bank*, p. 192.

79. Allen, *Biosphere 2*, p. 8.

80. Allen, *Biosphere 2*, p. 71.

81. Allen and Nelson, *Space Biospheres*, p. 3.

82. Allen and Nelson, *Space Biospheres*, p. 3.

83. Allen and Nelson, *Space Biospheres*, p. 33. See Allen, *Biosphere 2*, pp. 33–49 on 'biospherics'; pp. 59–71 on 'biotechnics' and 'ecotechnics'.

84. Allen, *Biosphere 2*, p. 60.

85. Zabel, Hawes, Stuart and Marino, 'Construction and Engineering of a Created Environment', p. 55.

86. Allen, *Biosphere 2*, 'Cybernetics', pp. 115–25, quotes p. 115, 125.

87. Allen, *Biosphere 2*, ch. 8: 'The Assembly', pp. 101 ff., on p. 109.

88. Allen, *Biosphere 2*, p. 33.

89. Allen, *Biosphere 2*, p. 106.

90. Allen, *Biosphere 2*, p. 73.

91. Allen, *Biosphere 2*, p. 10.

92. See Gentry and Liptak, *The Glass Ark*. This book was written for youngsters from age eight to twelve but is enlightening also for adults regarding the technological optimism involved in the project.

93. Allen, *Biosphere 2*, 'Diversity', pp. 48–9. For a criticism of a 'terraformed' nature see T. W. Luke, 'Environmental Emulations: Terraforming Technologies and the Tourist Trade at Biosphere 2', in T. W. Luke, *Ecocritique: Contesting the Politics of Nature, Economy, and Culture* (Minneapolis and London: University of Minnesota Press,

1997), pp. 95–114.

94. Allen, *Biosphere 2*, p. 101.

95. Allen, *Biosphere 2*, p. 35.

96. Allen, *Biosphere 2*, p. 53.

97. Allen, *Biosphere 2*, p. 35 (original emphasis).

98. Allen, *Biosphere 2*, p. 36.

99. Allen, *Biosphere 2*, p. 36.

100. Allen, *Biosphere 2*, p. 35.

101. Gentry and Liptak, *The Glass Ark*, p. 37.

102. Allen, *Biosphere 2*, p. 82.

103. Gentry and Liptak, *The Glass Ark*, p. 37. However, Biosphere 2 also had to deal with 'stowaways' not deliberately invited into the biosphere, particularly mice. Alling and Nelson, *Life Under Glass*, pp. 171–2.

104. Gentry and Liptak, *The Glass Ark*, pp. 38–9.

105. Gentry and Liptak, *The Glass Ark*, p. 39.

106. Allen and Nelson, *Space Biospheres*, p. 58.

107. Allen and Nelson, *Space Biospheres*, p. 58.

108. Allen, *Biosphere 2*, p. 130.

109. Allen, *Biosphere 2*. Biospherian Jane Poynter in fact centres her autobiographical report on Mission One on the human relationships and particularly on her relationship to her colleague Taber MacCullum. J. Poynter, *The Human Experiment: Two Years and Twenty Minutes inside Biosphere 2* (New York: Thunder's Mouth Press, 2006).

110. Allen, *Biosphere 2*, pp. 69–70.

111. Allen, *Biosphere 2*, p. 127. Compare 'Room for eight', Alling and Nelson, *Life Under Glass*, p. 11.

112. Allen, *Biosphere 2*, p. 127.

113. J. D. Bernal, *The World, The Flesh and The Devil: An Inquiry into the Future of the Three Enemies of the Rational Soul* (1929; London: Jonathan Cape, 1970), pp. 23, 25.

114. Gentry and Liptak, *The Glass Ark*, p. 82.

115. K. Kelly, *Out of Control: The New Biology of Machines, Social Systems and the Economic World* (Reading, MA: Addison-Wesley, 1994), p. 165, on Biosphere 2 pp. 128–49.

116. D. Sagan, *Biospheres: Metamorphosis of Planet Earth [Reproducing Planet Earth]* (New York: Bantam Books, 1990), p. 195, on Biosphere 2 pp. 190–5.

117. On 'creating a second nature in our image' see Rifkin, *Algeny*, p. 252; Sloterdijk, *Sphären*, p. 255. On culture becoming 'a veritable "second nature"' at the moment 'the modernization process is complete and nature is gone for good' see literary theorist Fredric R. Jameson, *Postmodernism, or, The Cultural Logic of Late Capitalism* (1991; Durham, NC: Duke University Press 1995), p. ix.

118. See Luke, 'Environmental Emulations', p. 106 on a critical perspective on the 'Commodification of Ecology'.

119. *Total Recall* (USA: TriStar Pictures, 1990), directed by Paul Verhoeven, starring Arnold Schwarzenegger, Sharon Stone and Rachel Tricotin. The plot centres on the struggle over control of the federal colony of Mars in the year 2084. The earth's northern block has colonized Mars to wage a war against the numerically superior southern block. War efforts depend on a constant supply of the Martian ore of 'terbinium'. The mine is managed from a human habitat built deeply into the red rocks of Mars to shield humans from radiation. Sealed glass domes have been erected to protect life from the vacuum outside. Yet, under a corrupt regime, a socially and spatially segregated society has

developed. The upper, visible stratum comprises weathly space tourists, while the lower stratum consists of deviant, mutated and rebellious groups who manage the nightclub district of the colony. Monopolizing the circulation of air is a means to secure power to the governnor, who can turn off the life supporting ventilation systems and depressurize the respective sectors at will. Notably, the remake of 2012, while maintaining the plot, abandoned Mars and situated the conflict on Earth.

120. Among the problems were temperature regulation, maintenance of the water cycle, and atmospheric composition. Species loss caused additional work. Kelly, *Out of Control*; see also a one-page note of correspondence by Eugene Odum in *Nature*, in which Odum comments on the 'extremely high cost of providing nature's free life support services with the use of non-renewable fossil fuels' as well as on the monetary costs of the project in general. E. P. Odum, 'Cost of Living in Domed Cities', *Nature*, 382 (1996), p. 18. The monthly electricity bills the crew members would have to face if they had to pay for utilities at the US residential rates would be more than $150,000. 'At anywhere near this cost, very few of the billions of people on Earth could afford to live in domed cities', Odum concludes.

121. W. J. Mitsch, 'Preface: Biosphere 2 – The Special Issue', in Marino and Odum, *Biosphere 2*, pp. 1–2, quote p. 1. See also Kelly, *Out of Control*, ch. 9: 'Pop Goes the Biosphere', pp. 150–65.

122. Mitsch, 'Preface: Biosphere 2', p. 1. The actual costs of Biosphere 2 are as contended as everything else about the project. Poynter in *The Human Experiment* speaks of $250 million (p. 5); the *New Scientist* mentions $150 million; see R. Lewin, 'Living in a Bubble', *New Scientist*, 4 April 1992, pp. 12–13; S. Veggeberg, 'Escape from Biosphere 2', *New Scientist*, 25 September 1993, pp. 22–4.

123. P. Warshall, 'Lessons from Biosphere 2: Ecodesign, Surprises, and the Humility of Gaian Thought', *Whole Earth Review*, 89 (1996), pp. 22–7, quote p. 27.

124. Alling and Nelson, *Life Under Glass*, p. 233.

125. Kelly gave a sceptical account in the *Whole Earth Review* while the experiment was still in full swing. As early as 1992, he reported on erupting controversy among the crewmembers. He also recounts that one Biospherian cut her finger off in a machine and was let out through the airlock for medical treatment, only to bring a bag of supplies back in after surgery. Kelly, 'Biosphere 2 at One', p. 90.

126. Mark Nelson and Bill Dempster, personal communication in 2007.

127. Mitsch, 'Preface: Biosphere 2', p. 1.

128. See J. Allen, M. Nelson and A. Alling, 'The Legacy of Biosphere 2 for the Study of Biospherics and Closed Ecological Systems', *Advances in Space Research: The Official Journal of the Committee on Space Research* (COSPAR), 31 (2003), pp. 1629–39. Biospherians Jane Poynter and Taber MacCallum, together with Grant A. Anderson filed United States Patent 5865141 for 'Stable and reproducible sealed complex ecosystems' in 1996, claiming an 'apparatus and method for establishing a sealed ecological system that is self-sustaining, remains in dynamic equilibrium over successive generations of organisms'.

129. Marino and Odum, 'Biosphere 2: Introduction and Research Progress', p. 13.

130. Luke, 'Environmental Emulations', pp. 101, 109; T. W. Luke, 'Biospheres and Technospheres: Moving from Ecology to Hyperecology with the New Class', in T. W. Luke, *Capitalism, Democracy, and Ecology: Departing from Marx* (Urbana and Chicago, IL: University of Illinois Press, 1999), pp. 59–87. On the relation of the simulation and the simulated in Biosphere 2 see N. K. Hayles, 'Simulated Nature and Natural Simulations:

Rethinking the Relation Between the Beholder and the World', in W. Cronon (ed.), *Uncommon Ground: Rethinking the Human Place in Nature* (New York and London: W.W. Norton, 1996), pp. 409–25; J. Baudrillard, *Simulacra and Simulation* (Ann Arbor, MI: University of Michigan Press, 1994) [orig. French *Simulacres et Simulation*, Paris 1981].

131. Odum, 'Cost of Living in Domed Cities', p. 18.

132. Poynter, *The Human Experiment*, p. vii. Hers is a telling account of foibles and conflicts among the group. See ch. 12: 'Cabin Fever', pp. 167 ff.; ch. 18: 'Dysfunctional Family', pp. 261 ff.

133. J. Baudrillard, 'Überleben und Unsterblichkeit', in D. Kamper and C. Wulf (eds), *Anthropologie nach dem Tode des Menschen: Vervollkommnung und Unverbesser- lichkeit* (Frankfurt a. M: Suhrkamp, 1994), pp. 335–54, on pp. 338, 343, 347. On the same note see also J. Baudrillard, 'Maleficent Ecology', in J. Baudrillard, *The Illusion of the End* (Cambridge: Polity Press, 1994), pp. 78–88 [orig. French *L'illusion de la fin*, 1992]; J. Baudrillard, 'Biosphère II', in A.-M. Eyssartel and B. Rochette (eds), *Des mondes inventés: les parcs à thème* ('Penser l'espace') (Paris: Éditions de la Vilette, 1992), pp. 127–30.

134. In *Logan's Run*, memory of the past world comes to the domed people from the out- side. Recollections are represented by a human being – Peter Ustinov is the Old Man – as well as by the specific site and architecture of Washington, DC. The overgrown buildings along the National Mall – the Washington Monument, the Lincoln Memo- rial, the Capitol – serve as stony materialized memories, as do the Capitol's paintings, books and statues.

135. Poynter, *The Human Experiment*, ch. 23: 'April Fools', pp. 323–9, on sprouting lawsuits regarding allegations of mismanagement and vandalism. In 1996 the facility was placed under the management of Columbia University in New York and used as a research laboratory. In 2003 the University of Arizona assumed control. In June 2007 the jour- nal *Nature* reported that Biosphere 2 and the surrounding land were sold for US $50 million to CDO Ranching & Development. In an ironic twist the housing develop- ment company intended to build 1,500 suburban family homes around the complex. Dalton, Rex, 'Biosphere 2 Finds a Buyer', *Nature*, 447 (2007), p. 759. In 2012, news went out that Biosphere 2 would serve climate change research, as a laboratory to study the Earth's water cycles. 'Big Science at Biosphere 2', *The University of Arizona* website, 24 April 2014, at http://www.arizona.edu/big-science-biosphere-2 [accessd 28 August 2014].

136. The most spectacular and sublime images of Earth seem to be those that are devoid of humans. The film images Sol sees on his deathbed when undergoing socially advised voluntary euthanasia in the movie *Soylent Green* present such a view on nature's bygone and perhaps future beauty.

137. Foucault lists heterotopias of 'compensation', sites which counterbalance disorder and messiness, like the colony, and heterotopias of 'illusion', exclusive and exclusionary sites of irregularity and departure. Foucault, 'Of Other Spaces', p. 27. Foucault also names 'crisis heterotopias', privileged, sacred, or forbidden places reserved for individuals who are, in relation to the human environment in which they live, in a state of crisis; p. 24.

138. Referring to Carl Schmitt's influential work on sovereignty, Giorgio Agamben devel- oped his theory of the exception on the case of the concentration camp or death camp. G. Agamben, *State of Exception* (Chicago, IL, and London: University of Chicago Press, 2005); G, Agamben, *Homo Sacer: Sovereign Power and Bare Life* (Stanford, CA:

Stanford University Press, 1995). For an argument 'against ecological sovereignty' on the global scale of environmental politics see M. Smith, 'Against Ecological Sovereignty: Agamben, Politics and Globalisation', *Environmental Politics*, 18 (2009), pp. 99–116. Smith maintains that the declaration of a state of global ecological emergency following a modernized Gaian reasoning will end in suppressing political liberties 'in the name of survival' rather than encourage ecological politics. The recognition of ecological crisis will go along with impositions of emergency measures and options for technological and military fixes (p. 110).

139. Latour, *We Have Never Been Modern*, pp. 13 ff.
140. Stewart Brand as quoted in Allen, *Biosphere 2*, p. 11. Brand introduced this edition of the *Whole Earth Catalog* with a statement of purpose claiming: 'We *are* as gods and might as well get used to it'.

6 Departure: The Habitats of Tomorrow

1. *Zardoz* (USA: 20th Century Fox, 1974), directed by John Boorman, starring Sean Connery, Charlotte Rampling and John Alderton.
2. M. Jäger and G. Kohn-Waechter, 'Materialien zur ökologischen Katastrophe: Das Verlassen der Erde' (4 Parts), Part I, *Kommune*, 1 (1993), pp. 33–8, quote p. 35 (translation is mine).
3. M. Foucault, *The History of Sexuality*, Vol. 1: *An Introduction* (New York, Pantheon Books, 1978) [orig. French *Histoire de la sexualité: La Volonté de Savoir*, Paris, Editions Gallimard 1976], p. 95.
4. Allen and Nelson, *Space Biospheres*, pp. 1, 59.
5. Allen, *Biosphere 2*, p. 147.
6. Allen and Nelson, *Space Biospheres*, p. 40.
7. Allen and Nelson, *Space Biospheres*, p. 52.
8. S. Hawking, 'How can the human race survive the next hundred years?' *Yahoo! Answers*, at http://answers.yahoo.com/question/?qid=20060704195516AAnrdOD [accessed 28 August 2014]. See also 'Hawking: Mankind has 1,000 years to escape Earth', *RT Question More*, at http://rt.com/news/earth-hawking-mankind-escape-702/ [accessed 28 August 2014].
9. P. R. Sahm, H. Rahmann, H. J. Blome and G. P. Thiele (eds), *Homo spaciens: Der Mensch im Kosmos* (Hamburg: Discorsi, 2005).
10. H. Blumenberg, 'Das Jahr 1969: Mondbezwingung und Umweltschutz', in H. Blumenberg, *Die Vollzähligkeit der Sterne* (Frankfurt: Suhrkamp, 2000), pp. 439–40.
11. The cartoon was released on 17 April 1980. It reappeared in *The Far Side*, by Gary Larson (the first collection of *The Far Side*), Kansas City/New York, Andrews and McMeel, A Universal Press Syndicate Company 1982 ©1982 by the Chronicle Publishing Company, p. 8.
12. Haraway, *Modest_Witness*, p. 169.
13. Haraway, *Modest_Witness*, p. 170.
14. Haraway, *Modest_Witness*, p. 170.
15. K. T. Litfin, 'The Global Gaze: Environmental Remote Sensing, Epistemic Authority, and the Territorial State', in M. Hewson and T. J. Sinclair (eds), *Approaches to Global Governance Theory* (Albany, NY: SUNY Press, 1999), pp. 73–96.
16. Google Lunar X Prize, at http://www.googlelunarxprize.org [accessed 28 August 2014].

17. J. Foust, 'Google's Moonshot', *Space Review*, 17 September 2007, at http://www.the spacereview.com/article/957/1 [accessed 28 August 2014].

18. J. Hanke, Director of Google Earth and Maps, 'Dive into the new Google Earth', February 2, 2009, The Official Google Blog, at http://googleblog.blogspot.com/2009/02/dive-into-new-google-earth.html [accessed 28 August 2014].

19. The first remote sensing satellite was put into space in 1972. NASA's new Blue Marble of 2012 was composed entirely of remote satellite data. See N. Wormbs, 'Eyes on the Ice: Satellite Remote Sensing and the Narratives of Visualized Data', in M. Christensen, A. Nilsson and N. Wormbs (eds), *Media and the Politics of Arctic Climate Change: When the Ice Breaks* (Basingstoke: Palgrave Macmillan, 2013).

20. S. Helmreich, 'From Spaceship Earth to Google Ocean: Planetary Icons, Indexes, and Infrastructures', *Social Research*, 78 (2011), pp. 1211–42; M. Wertheim, *The Pearly Gates of Cyberspace: A History of Space from Dante to the Internet* (New York and London: W. W. Norton, 1999), on the historical and epistemological movements from physical space to hyperspace and cyberspace.

21. A common project of Google Earth and the United Nations Environment Programme (UNEP) turned out a bestseller of spectacular 'before and after' pairs of satellite images: *One Planet, Many People: Atlas of Our Changing Environment* (Nairobi: UNEP, 2005).

22. SimEarth: The Living Planet, developed by Maxis Software Inc. in 1990.

23. 'Sim-Bio 2: Investigate Biosphere 2 by Modeling and Designing Experiments', *BioQuest Curriculum Consortium*, at http://bioquest.org/simbio2.html [accessed 28 August 2014].

24. Haraway, *Modest_Witness*, p. 290 fn 56.

25. 'Definition of Life Support Systems in the Context of the EOLSS', *Encyclopedia of Life Support Systems* (*EOLSS*), at http://www.eolss.net [accessed 28 August 2014].

26. 'Definition of Life Support Systems in the Context of the EOLSS'.

27. 'NASA Leadership and America's Future in Space'.

28. C. Kwa, 'Local Ecologies and Global Science: Discourses and Strategies of the International Geosphere-Biosphere Programme', *Social Studies of Science*, 35 (2005), pp. 923–50.

29. W. Sachs, 'One World', in W. Sachs (ed.), *The Development Dictionary: A Guide to Knowledge as Power* (London: Zed Books, 1992), pp. 102–15, on p. 108.

30. T. W. Luke, 'Environmentalism as Globalization from Above and Below: Can World Watchers Truly Represent the Earth?', in P. Hayden and C. el-Ojeili (eds), *Confronting Globalization: Humanity, Justice and the Renewal of Politics* (New York: Palgrave Macmillan, 2005), pp. 154–71. Luke objects that earth systems sciences aim to achieve optimality in a Benthamite sense. The British philosopher Jeremy Bentham's utilitarian imperative in the nineteenth century was to produce adequate wholes that promoted 'the greatest happiness of the greatest number'. J. Bentham, *The Works of Jeremy Bentham*, 11 vols (Edinburgh: William Tait, 1838–43), vol. 10, p. 142. Structurally, the organized way of assessing the environment in total resembles Bentham's panoptical view that Foucault described as the organizing principle of modern liberal societies. M. Foucault, *Discipline and Punish: The Birth of the Prison* (New York: Vintage Books, 1995) [orig. French *Surveiller et punir*, Paris, Gallimard, 1975].

31. M. Wackernagel and W. Rees, *Our Ecological Footprint: Reducing Human Impact on the Earth* (Gabriola Island, BC: New Society Publishers, 1996). The authors originally called the footprint 'appropriated carrying capacity'.

32. W. van Dieren (ed.), *Taking Nature into Account: A Report to the Club of Rome* (New York: Springer, 1995). S. Höhler and R. Ziegler (eds), *Nature's Accountability*, theme issue *Science as Culture*, 19 (2010); L. Haller, S. Höhler and A. Westermann (eds), *Rechnen mit der Natur: Ökonomische Kalküle um Ressourcen*, theme issue *Beiträge zur Wissenschaftsgeschichte*, 37 (2014).

33. G. C. Daily and K. Ellison, *The New Economy of Nature: The Quest to Make Conservation Profitable* (Washington, DC: Island Press/Shearwater Books, 2002); G. C. Daily (ed.), *Nature's Services: Societal Dependence on Natural Ecosystems* (Washington, DC: Island Press, 1997).

34. R. Costanza et al., 'The Value of the World's Ecosystem Services and Natural Capital', *Nature*, 387 (1997), pp. 253–60; for an update R. Costanza et al., 'Changes in the Global Value of Ecosystem Services', *Global Evironmental Change*, 26 (2014), pp. 152–8.

35. T. Saraiva, 'Breeding Europe: Crop Diversity, Gene Banks, and Commoners', in N. Disco and E. Kranakis (eds), *Cosmopolitan Commons: Sharing Resources and Risks Across Borders* (Cambridge, MA: MIT Press, 2013), pp. 185–212.

36. T. C. Boyle, *A Friend of the Earth* (2000; London: Bloomsbury, 2001), p. 248.

37. N. Klein, *The Shock Doctrine: The Rise of Disaster Capitalism* (New York: Henry Holt & Co, 2007).

38. *The World. Residences at Sea*, at http://www.aboardtheworld.com [accessed 28 August 2014].

39. *WALL•E* (USA: Walt Disney Pictures/Pixar Animation Studios, 2008), directed by Andrew Stanton. The opening sequence shows an image of planet earth as seen from outer space before zooming in through a thick orb of space debris to earthly wastelands and polluted industrial sites, ultimately turning to a demolished US east coast city skyline resembling the island of Manhattan and its dusty deserted streets where WALL-E is absorbed in his task.

40. P. J. Crutzen and E. F. Stoermer, 'The "Anthropocene "', *Global Change Newsletter*, 41 (2000), pp. 17–18; P. Crutzen et al., *Das Raumschiff Erde hat keinen Notausgang* (Berlin: Suhrkamp, 2011).

41. D. Harvey, *A Brief History of Neoliberalism* (Oxford: Oxford University Press, 2005), focuses on the years 1978 to 1980 as a revolutionary turning point of neoliberalism, marked by the liberalization of China's economy by Deng Xiaoping, Margaret Thatcher becoming Prime Minster of the United Kingdom in 1979, and the election of Ronald Reagan as US President in 1980.

42. See Z. Bauman, *Intimations of Postmodernity* (London and New York: Routledge, 1992); J.-F. Lyotard, *The Postmodern Condition: A Report on Knowledge* (Manchester: Manchester University Press, 1984) [orig. French: *La condition postmoderne: rapport sur le savoir*, Paris, Minuit 1979]. The American political scientist Francis Fukuyama held an antithetical view in 1992 on the end of history as the ultimate teleological dissolution of antagonism after the collapse of authoritarian regimes. F. Fukuyama, *The End of History and the Last Man* (New York and Toronto: Maxwell Macmillan International, 1992).

43. Thomas Kuhn originated the concept of 'normal science' in S. T. Kuhn, *The Structure of Scientific Revolutions* (Chicago, IL: University of Chicago Press, 1962). For 'post-normal science' see S. O. Funtowicz and J. R. Ravetz, 'Science for the Post-Normal Age', *Futures*, 25 (1993), pp. 739–55; F. Luks, 'Post-normal Science and the Rhetoric of Inquiry: Deconstructing Normal Science?', *Futures*, 31 (1999), pp. 705–19.

44. M. Gibbons, C. Limoges, H. Nowotny, S. Schwartzman, P. Scott and M. Trow, *The New Production of Knowledge: The Dynamics of Science and Research in Contemporary Societies* (London: Sage, 1994).

45. Notorious catalyst of the Science Wars was the 'Sokal Affair', the American physics professor Alan Sokal's exposure of what he called the 'fashionable nonsense' of leftist French postmodernists. Sokal's hoax article – he called it a parody – in the journal *Social Text* in 1996 was only the tip of the iceberg representing the struggle over the sovereignty of bestowing meaning on the world 'out there'. A. D. Sokal and J. Bricmont, *Fashionable Nonsense: Postmodern Intellectuals' Abuse of Science* (New York: Picador, 1998); P. R. Gross and N. Levitt, *Higher Superstition: The Academic Left and its Quarrels with Science* (Baltimore, MD: Johns Hopkins University Press, 1998).

46. B. Latour, 'Why has Critique Run out of Steam? From Matters of Fact to Matters of Concern', *Critical Inquiry*, 30 (2004), pp. 225–48.

47. W. Sachs, 'Zählen oder Erzählen? Natur- und geisteswissenschaftliche Argumente in der Studie "Zukunftsfähiges Deutschland"', *Wechselwirkung*, 17 (1995), pp. 20–5.

48. Latour, *We Have Never Been Modern*, p. 128.

49. B. Latour, 'Von "Tatsachen" zu "Sachverhalten". Wie sollen die neuen kollektiven Experimente protokolliert werden?', in H. Schmidgen, P. Geimer and S. Dierig (eds), *Kultur im Experiment* (Berlin: Kadmos, 2004), pp. 17–36; M. Groß, H. Hoffmann-Riem and W. Krohn, *Realexperimente: Ökologische Gestaltungsprozesse in der Wissensgesellschaft* (Bielefeld: Transcript, 2005).

50. U. Beck, *Risk Society: Towards a New Modernity* (London: Sage, 1992) [orig. German *Risikogesellschaft: Auf dem Weg in eine andere Moderne* (Frankfurt: Suhrkamp, 1986)]; U. Beck, A. Giddens, and S. Lash, *Reflexive Modernization: Politics, Tradition and Aesthetics in the Modern Social Order* (Stanford, CA: Stanford University Press, 1994); U. Beck, *World Risk Society* (Malden, MA: Polity Press, 1999) [orig. German *Weltrisikogesellschaft, Weltöffentlichkeit und globale Subpolitik* (Wien: Picus, 1997)].

51. *Our Common Future*, World Commission on Environment and Development (Chairman Gro Harlem Brundtland) (Oxford and New York: Oxford University Press, 1987), p. 43.

52. *Our Common Future*, p. 43.

53. *Our Common Future*, p. 27.

54. *Our Common Future*, p. 27. On the common 'we' in global environmental discourse that bypasses the terrain of differentiated politics see Taylor and Buttel, 'How Do We Know We Have Global Environmental Problems?'; F. H. Tenbruck, 'The Dream of a Secular Ecumene: The Meaning and Limits of Policies of Development', in M. Featherstone (ed.), *Global Culture: Nationalism, Globalization and Modernity* (London: Sage, 1990), pp. 193–206.

55. *Our Common Future*, p. 49. 'Beyond the Limits' was the title of the twenty-year update to *The Limits to Growth* in 1992: D. Meadows, D. Meadows and J. Randers, *Beyond the Limits* (Post Mills, Vermont: Chelsea Green Publishing, 1992).

56. See, for example, A. P. J. Mol, *Globalization and Environmental Reform: The Ecological Modernization of the Global Economy* (Cambridge, MA: MIT Press, 2001). For a critical reflection see M. A. Hajer, *The Politics of Environmental Discourse: Ecological Modernization and the Policy Process* (Oxford: Oxford University Press, 1995).

57. W. Sachs, 'Is small still beautiful? E. F. Schumacher im Zeitalter der grenzenlosen Mega-Ökonomie', in *Re-Vision: Nachdenken über ökologische Vordenker* (*Politische Ökologie*, 24 (2006)), pp. 24–6.

58. M. Levinson, *The Box: How the Shipping Container Made the World Smaller and the World Economy Bigger* (Princeton, NJ: Princeton University Press, 2006).

59. US and World Population Clock. United States Census Bureau, at http://www.census. gov/popclock/ [accessed 28 August 2014].

60. In 1967, when President Lyndon B. Johnson welcomed the 200-millionth citizen, about three quarters of the population were white; in 2006, about 56 percent were white. See 'From 200 to 300 million, how we've changed', *NBC News*, 17 October 2006, at http://www.msnbc.msn.com/id/15291977 [accessed 28 August 2014].

61. D. G. McNeil, Jr, 'Population "Bomb" May Just Go "Pop!" ' Low Birthrates Mark a Change', *New York Times*, 6 September 2004; M. Connelly, 'Population Control is History: New Perspectives on the International Campaign to Limit Population Growth', *Comparative Studies in Society and History*, 45 (2003), pp. 122–47.

62. See, for instance, W. Lutz, B. C. O'Neill and S. Scherbov, 'Europe's Population at a Turning Point', *Science*, 299 (2003), pp. 1991–92; S. P. Morgan, 'Is Low Fertility a Twenty-First-Century Demographic Crisis?', *Demography*, 40 (2003), pp. 589–603. See also J. E. Ehmer, J. Ehrhardt, and M. Kohli (eds), *Fertility in the History of the 20th Century: Trends, Theories, Policies, Discourses*, special issue *Historical Social Research/ Historische Sozialforschung*, 36 (2011).

63. P. D. James, *The Children of Men* (1992; London: Penguin, 1994), on pp. 10–11.

64. James, *The Children of Men*, p. 11.

65. James, *The Children of Men*, p. 281.

66. James, *The Children of Men*, p. 313.

67. James, *The Children of Men*, pp. 11–12.

68. STS scholarship has convincingly linked population studies across historical periods, pointing out that population studies often transport and reaffirm older eugenicist ideals: D. Kevles, 'Is the Past Prologue? Eugenics and the Human Genome Project', *Contention: Debates in Society, Culture, and Science*, 2 (1993), pp. 21–37; E. Ramsden, 'Confronting the Stigma of Eugenics: Genetics, Demography and the Problems of Population', *Social Studies of Science*, 39 (2009), pp. 853–84. Still, for the most part, due to the focus on the nation-state in population research, the human ecological assessments of populations developed in the 1960s and 1970s are not featured in these studies.

69. *Children of Men* (USA/GB: Universal 2006), based on the novel by P. D. James, directed by Alfonso Cuarón, starring Clive Owen, Julianne Moore and Michael Caine.

70. *The Possibility of Hope* (USA, Universal Studios, 2006) is a documentary film directed and produced by Alfonso Cuarón.

INDEX